Basic Information Systems Analysis and Design

Basic Information Systems Analysis and Design

Myrvin F. Chester
University of Wolverhampton

and

Avtar K. Athwall
University of Wolverhampton

THE McGRAW-HILL COMPANIES

London · Burr Ridge IL · New York · St Louis · San Francisco · Auckland
Bogotá · Caracas · Lisbon · Madrid · Mexico · Milan
Montreal · New Delhi · Panama · Paris · San Juan · São Paulo
Singapore · Sydney · Tokyo · Toronto

Published by McGraw-Hill Education
Shoppenhangers Road
Maidenhead
Berkshire
SL6 2QL
Telephone: 44 (0) 1628 502 500
Fax: 44 (0) 1628 770 224
Website: www.mcgraw-hill.co.uk

British Library Cataloguing in Publication Data
A catalogue record for this book is available from the British Library

ISBN: 0 07 709784 X

Library of Congress Cataloguing in Publication Data
The Library of Congress data for this book has been applied for

Acquisitions Editor: Conor Graham
Editorial Assistant: Sarah Douglas
Senior Marketing Manager: Jackie Harbor
Senior Production Manager: Max Elvey
New Media Developer: Doug Greenwood

Produced for Mcgraw-Hill by Steven Gardiner Ltd
Text design by Steven Gardiner Ltd
Printed and bound in Great Britain by Bell and Bain Ltd, Glasgow
Cover design by Hybert Design

Contents

Preface

Aims and objectives

This book is an introduction to the essential features of the analysis and design of information systems (IS). It is intended to be used as a textbook for University students at level one and two, and also for those on further education diploma courses and studying A' levels. Furthermore, the book is also aimed at those students who are converting from first degrees, not in computing or IS, to a Masters degree in these subjects.

It is also hoped by the authors that employees in information technology (IT) and IS departments in commercial organizations will find the book useful. Certainly, those beginning a career in these areas will find the book to be a useful handbook for, and reminder of, the basic tools and techniques of the profession.

The book is deliberately not scholarly or academic, being a guide, in easily recognizable language, to those basic techniques and practices that are essential to the knowledge of anyone who is beginning a study of information systems development (ISD).

The methodology of structured systems analysis and design (SSADM) version 4+ is used as the medium for discussing the modelling of ISs, present and proposed, and for relational data analysis (RDA or normalization). Object oriented techniques are also glanced at for the analysis of ISs, and an introduction to the analysis of requirements for ISs includes a brief exposition of soft systems methodology (SSM). Also, in order to describe the processes carried out in ISs, there is a presentation of the techniques of decision tables, decision trees, and Structured English. Furthermore, the book stresses the very important area for the budding systems analyst of writing reports and carrying out presentations.

Bridging the analysis of the current IS and the design of a new one, the book presents the various procedures of logicalization and RDA. These lead to the design of the data for the new system as well as that of its processing requirements. The way data is to be stored in the proposed computer system is next dealt with, followed by the design of screens and reports. The final parts of the book deal with the installation of the new system and its documentation.

Structure and content of the book

Each chapter's initial sections are essential reading if the objectives of the book are to be met. However, towards the end, a chapter often has sections on more advanced and thoughtful topics. These may be omitted and the basic aims of the book will still be covered, but they are considered to be useful for level two students, those at Masters level, and advanced students at all levels.

Also, dotted around the text, there are many hints, pitfalls, and tips to help the student in the use of the described techniques. Based on the writers' experiences of many years in teaching systems analysis and design, they will be very useful in assisting the reader to avoid the problems that lie in wait for the unwary or naive exponent of the science and art of ISs analysis and design.

Internet site

There is an Internet site associated with the book. For the student, it contains extra tutorials and worked examples. For the lecturer, the site also has suggested answers to the tutorials in the book.

The Internet site is to be found at:

http:\\www.mcgraw-hill.co.uk/textbooks/chester

Target readership

This book should be indispensable for the introductory levels of courses in computing, computer science, information technology, and ISs—including business ISs. It is envisaged that the book will be adopted as the essential text for University Level 1 systems analysis modules, Level 2 systems design, and conversion courses at Masters level. Similar courses being run in further education will also find the book very useful, as would those for A' level students.

Undergraduates at Level 1 or 2 should study all sections of the book except for those headed **PG TOPIC** (Post Graduate Topic). However, Level 3 (Final Year) students might also be expected to find those sections useful.

Conversion Masters students are assumed to have a first degree in a non-computing subject, so no assumptions are made about their previous knowledge of the topics covered. In this sense they are treated in the text as Level 1 and 2 undergraduates. However, Post Graduate students are expected to be able to be more reflective and evaluative than Level 1 and 2 undergraduates; so, having covered the standard sections of the text, they should study the **PG TOPIC** sections where they are available.

Acknowledgements

This book is dedicated to Alfred Waller, to whom our particular thanks are due and who originally encouraged us to write this book.

Many of the people with whom the authors have worked have made contributions to the modules from which the content of the book has been developed. Particularly worthy of mention are, in no particular order:

Anne Latham
Gary Moore
Graham Winwood
Helena Grealish
Richard McKenna
Wendy Davies
Wendy Evans

We thank them all.

Avtar K. Athwall
Myrvin F. Chester
Wolverhampton, 2002

Reader's guide

General

Chapter 1 is a general introduction to the subject of the book, and discusses the important terms to be used. Throughout the book, definitions labelled as *DOC* are from A *Dictionary of Computing*, 4th edition, published by the Oxford University Press in 1997. Definitions with no attribution are our own.

Chapter 2 deals with the way systems analysis and design fit into the systems development life cycle of an information system (IS), together with a discussion of the role of the systems analyst.

Chapter 3 looks in more detail at the several ways in which the requirements of a new IS may be discovered, and their incorporation into the catalogue of requirements.

Chapter 4 discusses the important topics of report writing and presentations.

Chapter 5 introduces the techniques of soft systems methodology, their use in the discovery of the requirements of an IS, and possible organizational and social implications.

Chapters 6 and 7 are the parts of the book dealing with the way the relationship between data in a system can be modelled using SSADM entity relationship diagrams. The latter chapter also includes a section on the way the relationships between data govern the way tables or files are designed for use within computerized ISs.

Chapters 8 and 9 look at the techniques for modelling the processes of a system and the way information flows into, out of, and between those processes; as well as where information may be stored. These chapters use the modelling technique of SSADM data flow diagrams.

In Chapter 10, there is a description of the SSADM techniques of entity life histories. It also contains a look at the object oriented technique of use-case diagrams.

Chapter 11 concerns three ways in which processes in ISs may be described more formally than by using ordinary English. It covers decision tables, decision trees, and Structured English.

The next chapter is intermediate between the gathering and modelling of information about the current system and the generation of designs for the new IS.

Chapter 12 describes the procedure of logicalization, which follows from the diagrams of the current IS and leads to modelling the design of a new system.

In Chapter 13 the book introduces the concepts of information systems design; what it is for, and what makes good design.

Chapter 14 is about relational data analysis (RDA) or normalization. The data in the current system can be analysed to discover the way each group of information interacts with others. Also, for the new system to be built, RDA is carried out to build the computer files (or tables) that the system will need to operate efficiently and correctly.

Chapter 15 is about the data requirements of a proposed new IS; while Chapter 16 looks at the processing requirements of that proposed system.

Chapter 17 concerns the storage of data in computerized ISs, and Chapter 18 relates to the design of user interfaces to such a system. It concerns the way data gets into the system and the production of information by that system in terms of screens and reports.

The last two chapters are about the last stages of IS development. Chapter 19 concerns the way a new system takes over from the old and other factors involved in the installation of the new system, and Chapter 20 is about the documentation of a computerized IS needed for a complete representation of the entire system.

Introduction

- The point of systems analysis and design
- Systems and information systems
- Analysis, systems analysis, and the systems analyst
- Systems design
- PG TOPIC: Quality added and quality chains

1.1 The point of systems analysis and design

For any human endeavour, apart from the trivial, to have some hope of being successful, it must pass through a small set of essential phases. In ordinary language, these are:

1 Find out what needs to be done
2 Plan what needs to be done
3 Carry out that plan.

Added to these, often omitted but essential if similar operations are to be carried out in the future, should be:

4 Evaluate what has been done.

For the development of a new computerized information system, which is definitely not a trivial activity, similar stages are also necessary. It is necessary to discover what is required of the system. The new system needs to be planned and designed; it has to be built and tested during the implementation of the design; and, usually after starting to use the system, the finished development process ought to be reviewed to see what lessons might be learned for the future.

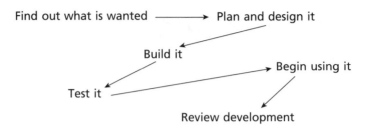

Finding out in detail what is needed, is carried out during the activities of information system analysis; the planning process is systems design; the building and testing procedures include detailed specification, development and testing of computer programs; and the review of the finished development is the system evaluation.

This book contains the basics for all of these activities except for the development of the individual programs; so there is a lot of ground to cover.

1.2   Systems and information systems

It is an illuminating exercise to look at the terminology being used. The term *system* is of much wider application than that represented by computer systems. Indeed, there is a whole field of study called *system theory*.[1] Everything that lives, moves, changes, or interacts with anything else is part of a system and itself constitutes a system.

All biological organisms are systems and made up of systems; and all machines are systems, as are their interactions with their environment. Also, particularly of interest for this book, all computerized information processors are systems within the greater systems of the businesses in which they run and the people with whom they interact.

The basic way in which a system of the most general kind can be envisaged is as a black box with inputs and outputs, which they take in and give out, shown in Figure 1.1.

feedback   return of part of the output of a system to the input as a means towards improved quality or self-correction of error (negative feedback) or, when in phase with the input signal, resulting in oscillation and instability (positive feedback)

Chambers

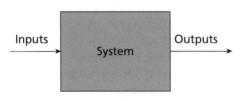

Figure 1.1   Black box model of a general system.

---

[1]   Perhaps the best book in this area is still *General System Theory* by Ludwig von Bertalanffy (1968).

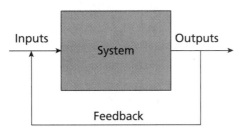

Figure 1.2   Black box model of a general system with feedback.

An input to such a general system can be anything from a very wide class. A biological system (or *biosystem*) takes in air, food, and water. A machine inputs raw materials and fuel; while the predominant input for information systems is raw data. System outputs can be waste matter, growth and muscular activity for biosystems; and noise, pollution, and finished materials for machinery. The aim of commercial computer systems is to output useful information in a variety of forms.

The simple model pictured in Figure 1.1 may be complicated a little to form Figure 1.2.

This model now includes the extremely important feature that is often known as a feedback loop.

All systems require such feedback to ensure their efficient and continuing functioning. Feedback looks at the output from the system and compares it with what the system is expected to produce. If the output is different from what is expected or required, the feedback loop passes that information back to the system to alter its inputs.

In biosystems, some of the output is the heat generated by the biological entity. If that heat is too great it presents a great danger and must be reduced. Biological feedback mechanisms detect excessive heat and make adjustments to the system, such as increased sweating and the reduction of activity, to cool the creature's body and reduce the threat. If the temperature of the body is too low, animals will increase their intake of food and move their position to increase the heat of the sun or a fire.

A machine usually incorporates a feature such as the steam engine's *governor*, which spins at a rate controlled by the speed of the engine. When the engine is rotating too fast for its settings, the governor automatically reduces the steam pressure thereby reducing the engine's function. If too slow, the governor increases the pressure and therefore the speed of the engine. When a system reaches the optimum conditions for what is expected of it, it is said to be in *homeostasis*.

For information systems (ISs), the feedback mechanism begins with the effect of its outputs, such as reports on paper and computer screens, on the people who look at them. If these outputs are not up to scratch and are not what those people expect to see, this should be reported to those who run or developed the system so

GIGO   Acronym for garbage in garbage out, signifying that a program working on incorrect data produces incorrect results.

DOC

that they can alter the inputs to the system or change the system itself. Often the poor state of the information coming out of an information system is a reflection of the poverty of the data that is entered into it. This is commonly heard phenomenon of *garbage in—garbage out.*

This input may need to be improved or filtered better so that, as far as possible, only high quality data gets in. However, the computer system itself may be in error. If so, it must be modified so that its output more closely matches what is expected of it.

Rarely is it sufficient to consider a system as one black box with its inputs and outputs. Systems are generally built-up of smaller systems—called *subsystems.* The opaque system box of Figure 1.2 can be opened up a little to show what it looks like inside. This model is shown in Figure 1.3.

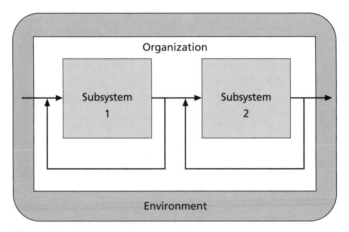

Figure 1.3   Model of subsystems of a system.

Here, we now have the overall system broken down into two subsystems. Note how the input and output to and from the environment (the outside of the system, with which it interacts) are the same for the overall system and the system broken into subsystems. What we have now is some more detail about precisely where the input goes to and the output comes from. The input goes into subsystem 1 and the output comes out of subsystem 2. Each subsystem has its own input and output, and its own feedback loop. Also, and very importantly, the output from subsystem 1 is the input for subsystem 2. This is the way subsystem 1 communicates with subsystem 2. In ISs, this is an *information flow* between the subsystems. Information flows from one subsystem to another.

In a business IS, information flows from one department to another, and, within departments, from one desk to another. This will be very important when we come to discuss data flows. Of course, in real systems, each subsystem may have several inputs and several outputs.

Subsystems may also need to be broken into even lower level systems. In systems analysis, this breaking down (*decomposition*) continues until no useful purpose is served by further analysis. Once the subsystem's function is simple enough to be clearly understood, decomposition is halted. Some subsystems will be quite complex and require several levels of decomposition, while others will be comparatively simple and require less to be fully described and understood.

ISs are, therefore, to be seen as a special case of systems in general, but sharing many of their basic characteristics. There have been many attempts to define what precisely is an IS. We include three here:

> *Information system*
> A grouping of people, objects and processes ... that provides information about the organization and its environment. It should be useful to members and clients of that organization.
>
> Provides procedures to record and make available information, concerning part of an organization, to assist organization related activities.
>
> (Flynn, 1998)
>
> A computer-based system with the defining characteristic that it provides information to users in one or more organizations.
>
> (DOC)

Note that two of these definitions do not mention computers at all. An IS need not be computerized and perhaps most such systems in the world are not automated. Certainly, before the advent of computers, there were still complex ISs in existence from ancient times until after the Second World War.

1.3 Analysis, systems analysis, and the systems analyst

The term *systems analysis* is now in common use, but its meaning is often obscured behind its supposed abstruse and technical nature. Even those who are earning a living from being systems analysts may not be aware of why their job title is what it is. A clue to its meaning was suggested above in the way systems were broken down into subsystems. This way of looking at the process is to be found in the definition of the verb *analyse*. From one dictionary, we have:

> *analyse*   or analyze, verb transitive[:] to resolve or separate a thing into *its elements or component parts; to ascertain those parts; ...*
>
> *Chambers CD*

So *analysis*, 'the action or process of analysing', concerns the separation of things into their elements and ascertaining what those elements are. Its opposite is *synthesis*, which is about building things up from simpler parts. (Chemical

Aside

The TV play by Alan Bennett, *A Visit by Miss Prothero*, is about a retired manager who spent a large part of his working life developing a very complicated paper-based business IS. He was so involved with this that, as his leaving present, the staff gave him a fancy version of the huge diagram that represented his system. Miss Prothero arrives and slowly lets him know that the entire thing has been altered and simplified by the manager who has replaced him. This effectively destroys the rest of the life of the original developer.

Aside

God's IS. The Powell and Pressburger wartime film *A Matter of Life and Death* contains a marvellous example of the non-computerized system that purportedly exists in Heaven to cope with the lives and deaths of human beings. By the way, a bug in the system means that it fails to cope with the death of David Niven, which causes the problems dealt with in the film.

synthesis, for instance, takes simpler chemicals and puts them together to form more complicated ones.) So *systems analysis*, properly *information systems analysis*, relates to the separation of ISs into their component parts—the breaking down of the process of a complicated information system into its simpler parts.

ISs, even (or especially) those that are paper-based and non-computerized, can be devilishly complicated. Often much too complicated for the likes of us to understand completely. The human brain may simply not be capable of coping with all its intricacies.

The objective of systems analysis is to break down all this complexity into simpler parts, and the interactions between those parts, which we hope we shall be able to understand.

Many people would question the wisdom of trying to understand a complex object by breaking it down into simpler parts. Something is lost, they would say, when a flower is pulled apart in order to attempt an understanding of how it works. There is much to commend this view. However, it must not be forgotten that ISs—paper-based or otherwise—were invented by people; and none of them were invented by God.

Being people-made, and often being built up from an originally simpler system or the combination of several simpler systems, we might expect them to be susceptible to the attempts of people to understand them through analytical simplification.

IS analysis, as dealt with in this book, is the analysis of a currently operating IS in order to understand it well enough to use that knowledge for the development of a new computerized system. The current system may be completely paper-based, or it may contain some computerization, or it may even be completely computerized. Whatever the state of the present system, it needs to be broken down into its component parts and the interactions between those parts so as to understand it as thoroughly as possible.

All this interest in the workings of the current system is because the requirements of the new computerized system are very often dependent on what already happens. IS personnel are frequently asked to provide a computerized version of the paper-based system that exists at the moment. However, in addition to the knowledge obtained from the analysis of the current system, extra requirements for the new system have to be obtained from those whose job it is to help the IS people to find out what is needed in the new system. These additional requirements can be split into two types. Users of the old system would very much like the new system to correct the errors in the old system—these are its *problems*. Also, they would like improvements to the old system that were deficient there— these are extra *requirements* for the new system.

The analyst models the present IS (often called the *current physical system*, because it exists now and is implemented through physical objects such as paper forms, people, telephones and such like) by producing diagrams of how the system works and the way its data is interrelated. The models for the working of the current system will be taught here as *data flow diagrams (DFDs)*. Any problems with this current system are noted, as are any additional functions that the user wants.

Before progressing to the design of a new system, the current system DFDs need to be divorced from the way the present system is physically implemented.

System problems Errors and mistakes in the current system that the users would like to have improved

System requirements Extra functions, not present in the current system, that the users would like to see in the new one.

The reason for this is that we are using the models of the current system as a basis for the new (or proposed) IS—which will probably be, in large part, computerized. We do not want to be hampered by the way the system is implemented at the moment. There should be no distractions, such as the way that certain functions of the present system are carried out in particular departments or by particular people. This may not be the case with the new system. Also, the fact that, at the moment, orders are written on paper order forms should not be allowed to detract from other possibilities such as the input of orders via the Internet. We step back from the current system and abstract away its physical aspects. By doing this we produce the *logical view* of the current system (the *current logical system DFDs*)— this procedure is called *logicalization*.

The data of the current physical system must not be ignored. The analysts must uncover all the information that is held in the current system—what is stored about customers and orders and such like. We need to find out every last scrap of data that the present system handles. Also, we have spoken of the way the systems analyst discovers the way the data is interrelated. This means the way that data about one aspect of the system is related to other data in the system.

For instance, we need to know how the information held about customers relates to the information held about customer orders. Is an order always placed by one customer, and only one customer? Or, perhaps, is there a way in which two or more customers can place one order? This seems to be fairly rare in systems that deal with such things, but it is by no means impossible. Bank accounts are a good example of the way more than one person can have one account. Furthermore, there is the question of whether a customer is allowed to place more than one order. This is much more likely, and probably fervently to be wished. However, it is not impossible to imagine companies who insist that a customer can only have one order from them. But, even then, over time, the system may well be expected to be able to cope with the same customer placing another order after the previous one has been completed.

This is dealt with by a modelling technique that relates one type of data to another—this will be taught as *logical data modelling*, also known as the production of the *entity relationship diagrams* or *ERDs* (usually only one ERD) for the system.

The *systems analyst* is, fairly obviously, someone who undertakes the systems analysis. However, the job title often covers much more than the analysis. In many companies, the analysts are also the system designers, taking the requirements for a new system through to its design and testing as well.

Perhaps it would be educational to discuss the characteristics that might be expected of a good systems analyst. Psychologically, it has been said that though good computer programmers are often introverts, systems analysts are more often extroverts.

It is certainly true that systems analysts need to interact well with other people in their company, for they need to communicate with them to find out what they want from their systems, and also to inform them what their systems will be able to accomplish. However, in today's business organization, no one is able to be too introverted. A programmer who is only able to talk to computers, is a programmer who is unable to take part in the teamwork that is essential in the development of quality ISs. Certainly, a systems analyst who cannot speak to, or

The systems analysts are responsible for identifying a set of requirements ... and producing a design. The design is then passed to the programmers, who are responsible for [the] actual implementation of the system.

DOC

Extroverts are outgoing folk; they like to mix with other people, enjoying personal interaction at work and at leisure.

Introverts, on the other hand, tend to keep to themselves. They dislike interacting with other people and are for instance happier, it is said, talking only to the computer through their programs.

be spoken to by, other people is not going to be very good at the job. There will be more of this in Chapter 2.

1.4   Systems design

Stage two, mentioned at the beginning of this chapter following finding out what needs to be done, is 'plan what needs to be done'. Once the analysts have discovered, with some degree of certainty, what is required of a new IS, they need to move to the development of a blueprint of the new system. There may be people with the job title *system* designer, but they are few and far between. The job of designing a new computer system is generally carried out by the systems analysts who discovered the requirements of that system.

The systems analysis has taken the DFDs of the current IS and developed DFDs of the current logical IS. The systems design uses this logical model and combines it with what is known about the additional requirements for the system, to produce a *proposed logical system*. This is a logical (non-physical) view of the way the new system should work. It is the basis of the design of that proposed new IS.

There is another, vitally important aspect of the proposed system that has yet to be considered—the data for the new system. As mentioned above, the systems analysis has found all the information being handled by the current IS, considered the way the data is interrelated, and drawn the ERDs to model that relationship. It is now possible, by using the requirements for the new system and the models of the data and functions of the current system, to work on the way the data for the new system can be held.

The premise of this book is that the data for the new system will be held on a computer, and held in a relational database. The technique for developing the way data should be held in such a database is called *relational data analysis (RDA)*—also known as *normalization*. In RDA we take what is known about the information of the current system, and, often combining it with what we know about extra data needed for the proposed system, develop the computer files (or *tables*) for the new system. These tables will have as little duplication of data as possible, and will be related in such a way as to allow a new computer system to get all the data it requires to carry out its functions. From these tables, a new ERD can be drawn for the proposed system.

1.5   PG TOPIC: Quality added and quality chains

Like the latter parts of other chapters, this part of the chapter is where we consider what has just been discussed in a little more detail—and more critically.

Those who are paying for computer systems—those we often refer to by the rather dismissive term *users*—may well ask what the point is of IS personnel spending so much time and money in activities such as analysis and design, rather than actually writing programs. This is a question not only posed by the naive user, but also the naive programmer. Many academics have dismissed this question pointing to the obviousness of having to carry out such activities in order to produce good systems. Compare, for instance, the opening section of this chapter. But, in fact, it is not such a silly question as they might suppose.

For manufacturing business, the only reason for carrying out any process on a product is that it will add value to that product. The *value added*, as it is called, must be greater than the cost expended on carrying out the process. For example, giving a product a coat of paint is only justifiable economically if the customer is willing to pay more for a painted product than for an unpainted one. However, for the development of a software-based product, the non-programming activities advocated by this book and many, many others, do not obviously add value to the product. This section of the chapter offers a justification for analysis and design based on the concept, not of value added, but of *quality added* and its related concept of *quality chains*.

1.5.1 Quality added

The building in of quality is a frequently expressed aim of manufacturing procedures. The building of computer systems has similar objectives to produce software that is of high quality. Often though, these quality procedures have little justification apart from the pious but dogmatic claim that 'obviously' the processes of requirements analysis, design diagrams, and testing are necessary to the high quality of computer systems.

Time and money are spent on procedures such as systems analysis and design, because software developers believe that without such processes the user will not receive a product of high quality. In manufacturing, processes carried out on an item throughout its production are justified only if the cost of carrying out that process is less than the value added to the end product. If these economics were applied to all the phases of software development, it could be difficult to justify why many of the early stages are carried out.

Although a particular phase or process carried out during software development may not obviously add value to the product, Chester and O'Brien (1997) argued that it could nevertheless be justified if it can be shown to be adding quality. If an economic justification were required, it can be argued that a product with more added quality will often be seen to be of greater value to a purchaser than a product of lower quality. The authors noted that:

> A product may promise to be of value, but quality ensures that the value to the customer really is built into the product. There is therefore a justification for all the non-manufacturing phases in the software development life cycle: these processes add quality, and this increased quality adds value.
>
> (Chester and O'Brien, 1997)

1.5.2 Quality chains

Metaphor can often be usefully employed in understanding difficult and complicated processes. The quality chain was introduced in Chester and O'Brien (1997)

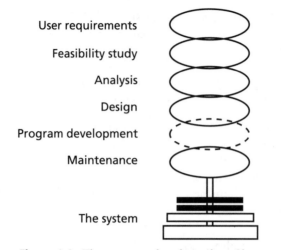

User requirements

Feasibility study

Analysis

Design

Program development

Maintenance

The system

Figure 1.4   The system quality chain (from Chester and O'Brien, 1997).

as an amusing and informative way of envisioning the way systems are dependant upon the quality of the various procedures that are undertaken during their development. The overall view of the quality chain of a system can be pictured by Figure 1.4.

Each stage or phase of information systems development (ISD) is shown as a link, and the strength of each link depends upon the quality added during that phase. The finished system hangs like a load from its quality chain. Some systems are a heavy load on their chains and some are lighter. Heavy loads are those that are considered to be important or large, while lighter loads are smaller and more trivial. The connections between a link and the links above and below it are also important. Chester and O'Brien (1997) write:

> This [quality] chain, each of its links, and the connections between those links, must be constructed carefully to provide adequate quality support for the running system.

The complexity in the area of program development—shown in the diagram as a dotted link—can be more clearly shown by Figure 1.5.

In Figure 1.5, each program has its own quality chain consisting of links for specification, design, coding, and testing. The quality of the whole program is dependent upon the quality added by these three stages of program development. The diagram above shows three programs that hang from the systems design, and from which and upon which the system eventually depends. Should one of these programs fail because of weak links, the system's overall quality will be affected, and the system load will be in danger of crashing in an unpredictable way.

The concept of quality added and the image of a quality chain from which the computer systems hangs should help in understanding why it is so important that the various phases of the life cycle are carried out—and carried out to a high

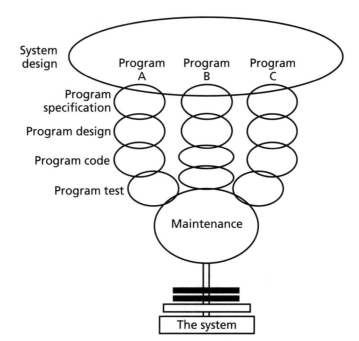

Figure 1.5 Quality chains in program development (from Chester and O'Brien, 1997).

standard. The systems development life cycle is discussed in more detail in the next chapter.

The systems development life cycle and the systems analyst

2.1 The systems development life cycle

The activities described in this book do not exist in isolation. They are a part of the whole set of procedures undertaken to produce a computerized information system (IS). This set of activities is commonly referred to as the *systems development life cycle (SDLC)*.

In ordinary terminology, the phrase *life cycle* is used to cover the birth, life, and death of an organism, or of an ecosystem. By analogy, the SDLC is:

> *System life cycle*
> The phases of development through which a computer-based system passes. The term is primarily used in the context of information systems.
>
> (DOC)

While another definition says:

> *Life cycle*
> The phases through which a development project passes from inception of the idea to completion of the product and its eventual decommissioning.
> (Bennett *et al.*, 1999)

For our purposes, either of these definitions will do. Like everything else, a computer system will come into existence, have an interesting life, and then it will die. (*Sic transit gloria mundi*[2]). Luckily for computer systems, and unlike us, once they have gone beyond their useful life, they can be replaced, we hope, by something better. Generally, the systems development life cycle is described in more detail than the simple born–live–die–replace view so far discussed. As shown in the previous chapter, a lot of things happen before the system is born. That is, before—as the jargon has it—it is made *live*. These activities of the life cycle are called the *phases*, or, sometimes, the *stages*, of the whole project. They are usually delineated as:

> **Stages in ISD**
>
> Scope and objectives
> Feasibility study
> Systems analysis
> Systems design
> Implementation: code, test and install
> Maintenance
> Review

2.1.1   Phased development

Because the system is developed in a number of stages or phases, this type of life cycle is referred to as phased development. A popular representation of such a phased development life cycle is the so-called Waterfall model, discussed later in Section 2.3.

2.2   The SDLC in detail

2.2.1   Scope and objectives

These constitute the early and very general initial order for the project—its initiation. Those who are responsible for paying for the project must begin the

---

[2]   So passes [all] worldly glory.

Feasibility study A study carried out prior to a development project in order that the proposed system is feasible and can serve a useful purpose.

DOC

process of its development by describing what they want, and what area of the business they want the computer people to investigate and improve. They may well also include details about the allowed budget of time, money and other resources. These outline directions are called the project's *terms of reference*.

2.2.2   Feasibility study

Following the directions in the terms of reference to begin investigating the project, IT staff will often undertake a *feasibility study*.

It is, in effect, a small systems analysis, that looks in a little detail at what is required and whether what is being asked for is likely to be possible given the technology available, and the time, money and staff that have been budgeted for in the terms of reference.

The feasibility study results in a *feasibility report* that sets out the findings of the study stating which parts of the initial request are feasible within the constraints of the terms of reference.

Systems analysis The analysis of the role of a proposed system and the identification of a set of requirements that the system should meet, and thus the starting point for system design

DOC

2.2.3   Systems analysis

The next few chapters discuss the basic techniques that can be used to carry out this phase of the life cycle.

2.2.4   System design

In this book, this activity follows the description of systems analysis techniques.

2.2.5   Implementation

That is, the implementation of the design in terms of the building or procuring of the software and hardware, setting them up to run properly, and getting the users from the old system to the new one – the system changeover.

It covers software (or program) development: the design, coding and testing of individual components of the software system; as well as link testing, system testing and acceptance testing. Added to this, the system must be installed so that it is up and running for the users to use.

This book excludes the development of the software but includes details of all the other aspects of implementation, including system changeover.

Installation of the system includes the setting up of all the hardware and software and the system changeover.

System design The activity of proceeding from an identified set of requirements for a system to a design that meets those requirements.

Implementation The activity of proceeding from a given design of a system to a working version ... of that system.

DOC

2.2.6   Changeover

That is, the procedure whereby the old IS is superseded by the new one. It is the birth of the new system. Before changeover the old system is running and, after changeover, the new one is running. To get from one state to the other is not an easy process (like human birth) and the book discusses several ways of carrying it out.

Pitfall

The word *implementation* has more than one meaning. Do not confuse our definition of the term with its other meaning, that of making a system *live*. This activity is here called *changeover*, and is discussed above.

System changeover The process of moving from the use of one IS to the use of another.

Pitfall

Forgetting about the maintenance phase can be very embarrassing. Many authorities have estimated that, over its lifetime, the cost of maintaining a system can be as much as the cost of developing it in the first place.

2.2.7   Maintenance and review

The new system has now been born and it is beginning its life of useful work. There are two very important aspects of information systems development that are often overlooked or otherwise skimped on because they occur after the exciting work of producing the new system.

Review is the activity of looking over what happened to generate the completed system. IS and business people should be looking for the good parts and the bad parts of the development process so that they can learn from what has happened in order to make future development easier, more efficient and more effective.

Maintenance takes place during the life of the new system. It refers to all the activities that take place during the period when the IS is running live. Changes will always be needed by any non-trivial system and the book looks at the reasons why this is so, and the various types of modifications that can be required.

2.3   Systems development life cycles

There are various methods of developing ISs. We have space to discuss briefly only a few.

2.3.1   Waterfall life cycle

This is considered to be the traditional way of developing computerized ISs. Each phase or stage is carried out to completion for the whole project. In its simple form, when one stage is finished, only then is the next stage begun. This life cycle is shown in Figure 2.1.

The implications of this model appear to be that the project moves from stage to stage in a logical manner without ever looking back to see if one stage had followed properly from the last, or if what is being generated now is what was wanted in the first place. We doubt if anyone seriously followed this life cycle in that way. However, a better representation of the Waterfall life cycle is that shown in Figure 2.2.

Figure 2.2 shows that iterations (repeated activities) can take place within the traditional Waterfall life cycle. Project phases can be revisited, checking of one stage against another may take place, as well as, checking that the present stage is in conformity with what was originally requested. This version is obviously more responsive to changes in the user requirements as the project progresses, and stresses the importance of *verification and validation*.

2.3.2   Verification and validation

This is a more refined way of talking about the checking of a particular stage or deliverable of the life cycle. These terms are often referred to as *V&V*.

Verification may be characterized as the checking of a deliverable to see if it follows logically (or sensibly) from its precursors—any stage of development must

Table 2.1 The phases and deliverables of the SDLC.

Phases	Deliverables*	Objective/Purpose
Scope and objectives	Terms of reference.	To identify and determine the problems. Establish the scope (boundary of the problem domain) and plan for a feasibility study.
Feasibility	Feasibility report.	To present a report to the management on the operational, technical and economical viability of the project to make a decision about the progression of the project.
Systems analysis	Requirements specification. Physical and logical models of the current environment.	To develop the requirements specification of the required system.
Systems design	Logical models of the required system. Physical models of the required system. Detailed design documentation.	To develop outline solution(s) and then provide a detailed specification of the required system.
Implementation:		
● Code	Code and program documentation.	To develop and deliver the working system according to the required specification.
● Test	Test plans and tests from unit to system and acceptance testing.	
● Install	Installation of hardware and software, user documentation and training. System changeover.	
Maintenance	Maintenance procedures.	To rectify problems encountered and to provide continuing support and maintenance.
Review	Evaluation report.	To establish that the user needs have been met.

* A *deliverable* is anything that is produced and presented (delivered) to those who need to see or use it.

be based upon the deliverables of the previous phase. If not, then something other than what was envisaged by the previous one will be produced.

Validation[3] is just as important. It can be seen as the checking of a deliverable against the original wishes as set out in the requirements of those who have sponsored the new IS.

---

[3]   Verification: 'Are we building the product right?'; Validation: 'Are we building the right product?' (Somerville, 1988)

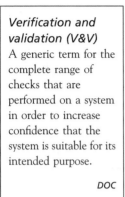

Verification and validation (V&V)
A generic term for the complete range of checks that are performed on a system in order to increase confidence that the system is suitable for its intended purpose.

DOC

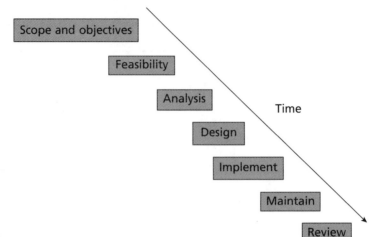

Figure 2.1   The simple Waterfall life cycle.

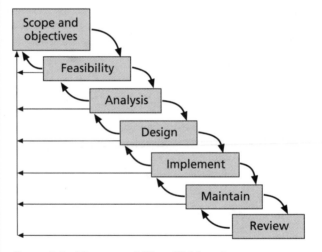

Figure 2.2   The practical Waterfall life cycle.

Pitfall

Logically minded people may believe that taking a series of steps, each following logically from the other, must result in a product that logically follows from the first step. In a logically perfect world this may be the case. In the real world, it is very easy to make small alterations while producing each deliverable, so that the finished product eventually is very different from what was originally intended. Computer personnel have often been accused of developing systems that no-one wants to use. So validation, which is keeping one's eye on the ball of the original requirements, is as essential as the verification that one stage follows logically from another.

The sort of life cycle exemplified by the Waterfall model is called *phased* and another of these phased life cycles is the V-model, which is not discussed here; but see Yeates and Cadell (1996).

The other major type of life cycle in use is called *evolutionary*. Evolutionary life cycles build the system from rough incomplete versions up to the final system. Here we look at the prototyping (and rapid application development) and spiral life cycles.

2.4   Other life cycles

So far, we have considered a commonly-used systems development life cycle; the full Waterfall life cycle is often promoted for large, important systems, but shorter varieties are often used for smaller systems and where speed is required in development. Other, rather different life cycles are also suggested and used. A few of these varieties are listed below.

2.4.1   Prototyping

A *prototype* is an early version of the final system and *prototyping* produces a preliminary version of the required system that can be reviewed by end-users. After review, the prototype is added to and altered to produce another version closer to the one that is wanted.

The first version is called the *first cut* prototype. The user reviews this and asks for improvements so the developer can go away and produce a bigger, better version. This is reviewed in turn, and subsequently improved. This process continues until the user is satisfied with the final version.

It is essential in prototyping to chop up the project into manageable chunks and that each cut is built rapidly. The user is not then kept waiting for a long time between reviews. It is an iterative process where users can suggest changes and modifications before further development takes place.

Figure 2.3 gives a diagram of prototyping process.

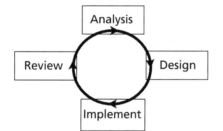

Figure 2.3   Prototyping process.

2.4.2   Types of prototyping

The development of mock-up IS prototypes ('*mock-up prototyping*') has been in use at least as long as online systems have been available. This type of prototyping is used to elicit from the user what screens would be acceptable in their content and aesthetic appeal, and the agreed screen layouts can then be used as templates for the input–output parts of the systems design. The mock-ups are used as guides and do not form any part of the finished system.

In an intermediate version, prototypes can be evolved with more functions than would be found in a simple screen mock-up. One of the prototyping methods, '*prototyping methodology*' (Lantz, 1986), uses prototyping tools such as fourth generation languages (4GLs) to develop a prototype system that not only has screens that look like those of the final system, but also appears to function, in many ways, like the final system. The prototype is modified after exposing it to the end-user, and can be used as a method of discovering more precisely what the user wants. However, in Lantz's methodology, the final system is produced from the user requirements using a 3GL.

The final type of prototyping discussed here ('*prototype as system*') is what might be thought to be the logical extension of the previous method. If the prototype can be built so as to have the same screens as the final version and also so as to have all the functions of the final version, then, with a little more work, the prototype can become the final version of the system. In this procedure, the prototype is developed with a view to becoming the final system, perhaps using third generation languages (3GLs), but very often using 4GLs. Beynon-Davies *et al.* (1995) call this method the '*rapid prototyping*', which 'covers the whole of the life-cycle albeit in a compressed period of time', of an 'incremental prototype . . . which, by incremental refinement, will form the whole of or part of a delivered system'.

2.4.3   Rapid Application Development (RAD)

RAD has been popularized and expounded by Martin (1991); it endeavours to be a complete method of developing computer systems quickly and effectively. Martin's book *Rapid Application Development* sets out all the steps, tools and techniques that should be used by RAD developers, including the way RAD IT development teams should be organized.

The RAD life cycle has four phases: the requirements planning phase; the user design phase; the construction phase; and the cutover phase. The requirements planning phase 'requires that high-level or knowledgeable end users determine what the functions of the system should be . . . with the guidance of I.S. professionals' (Martin, 1991, p. 81). The user design phase 'requires that end users participate strongly in the nontechnical design of the system' (Martin, 1991).

It uses integrated computer aided software engineering (I-CASE) tools and prototyping to help the users and IT personnel create the CASE design specifications. In the construction phase, 'I.S. professionals do the detailed design and code generation of one transaction after another, using the I-CASE toolset' (Martin, 1991, p. 84). In partial support of this, Moreton and Chester (1997, p. 99) note that:

> No full I-CASE tools yet exist, but when they do, they will cover the entire software life cycle, including project management, and will manage all information for the entire applications portfolio (including data models, data dictionary, databases, activity models, and so on).

Prototyping is used, so exposing each transaction to the end-users, and modifying it until it meets their requirements exactly. The cutover phase requires, *inter alia*, 'comprehensive testing . . . and operation in parallel with the previous system until the new system settles in' (Martin, 1991, p. 86). In the prototyping part of the construction phase, Martin is adamant that 'prototype[s] must be part of the evolving system . . . they should be built *with* the final development tool so that they pass directly from the User Design Phase to the Construction Phase' (Martin, 1991, p. 172).

In this way, RAD is an example of the 'prototype as system' method. The tools used may vary but they form an integrated CASE toolset; for instance, a screen painter, a 4GL, and a code generator, or a package such as the Unisys LINC II system generator that produces Common business oriented language (COBOL) code which gives 'good machine performance' (Martin, 1991, p. 177). Prototyping is also recommended for the user requirement and user design phases of the life cycle using the I-CASE toolset.

The benefits claimed for RAD are the rapid delivery of systems at a reduced cost compared to other methodologies. The use of prototyping is promoted because, 'When a prototype is reviewed seriously by end users they almost always change something … [it] does much to solve the problems of inadequate communication between designers and users' (Martin, 1991, p. 172). The use of a code generator as part of the I-CASE toolset produces code that 'should be free from coding errors (though it often has design errors). It can be run immediately, then adjusted as desired' (Martin, 1991, p. 84). Martin produces figures that suggest development improvements for companies that have used methodologies not unlike RAD.

2.4.4   Spiral method

The spiral model includes the best features of both the classic Waterfall SDLC and the prototyping approach. It is an evolutionary (iterative) systems development life cycle model developed by Boehm (1988). It was developed in recognition of the fact that systems development projects tend to repeat the stages of analysis, design and code as part of the prototyping process; and it also incorporates risk assessment.

The model is closely related to RAD, as it implies iterative development with a review possible after each iteration or spiral. This corresponds to the production of one prototype, incremental version, or cut.

Figure 2.4 shows a diagram of the spiral method.

Each spiral consists of four main activities:

- Plan next cycle: set the project objectives; define alternatives; further planning on the next spiral.
- Analyse the risk: analyse alternatives and identify and provide resolutions for the risks.
- Develop the system: design, code and test the system, in increments.
- Evaluate the system: the user evaluation of each spiral and then the final product.

2.5   The role of the systems analyst

Systems analyst is the usual job title for the member of staff carrying out the analysis and design of an IS. It is instructional to consider what function this

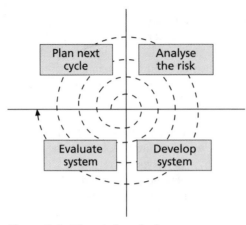

Figure 2.4 The spiral method.

person carries out in an organization and what sort of personality is required to make a good systems analyst.

2.5.1   A catalyst for change

The systems analyst has the important function of allowing the organization to change the way it carries out its IS activities. This role can be termed an *agent* or *a catalyst for change*. In chemistry, a catalyst is a chemical that affects a chemical reaction, to allow it to take place or to speed it up. In ordinary English, a person can also be referred to by this word.

Analysts use their professional skills to fashion the change from an old IS to a new one, or perhaps to allow the company to install a completely new one. Various personal characteristics may be considered to be necessary to carry out such a job effectively. A few of these are listed below.

> *catalyst* ... a person who causes or promotes change by their presence in a situation or their input into it (figurative).
>
> *Chambers CD*

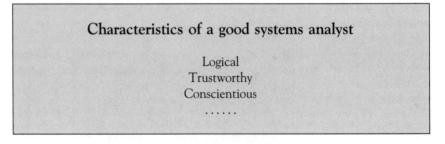

Characteristics of a good systems analyst

Logical
Trustworthy
Conscientious

. . . . . .

You should be able to think of several more of these.

2.6   SSADM

These initials stand for *structured systems analysis and design method (or methodology)*.

> **SSADM** The standard UK government analysis and design methodology ... SSADM covers the data (information), processing (functions), and events (logical) views of a system.
>
> *DOC*

The method is looked after by a committee and has undergone many revisions and additions; the version used in this book is version 4+.

This book does not aim to be an exhaustive exposition of SSADM or its uses; but it uses those parts of the methodology[4] that we are going to find useful in our descriptions of the basics of IS analysis and design. There are many texts dealing in much more detail with the methodology, such as Goodland and Slater (1995) and by our friend Malcolm Eva (1995). The reader is directed to these for more information on the complete methodology.

2.7   Other methodologies

SSADM is in common use in the UK, and indeed is essential for any company that expects to carry out IS work for the UK government. However, it is not as commonly used in other parts of the world. In the US particularly the most common methodology is probably that of Yourdon (1976), and that set out by such writers as De Marco (1979). The work of Gane and Sarson (1979) is also seen in use as *structured systems analysis (SSA)*. Although differing in detail, with very little study it ought to be possible for the student who understands the SSADM terminology and diagramming techniques taught in this book to convert fairly easily to these other two methods.

SSADM is commonly referred to as a *hard* methodology. This is to distinguish it, and its fellows, from the soft methods exemplified by soft systems methodology (SSM). SSADM is also a structured methodology, but there are methods that are not oriented to structures but are object oriented. SSM is discussed in Chapter 5, and some object oriented techniques will be considered briefly in Chapter 10.

In other parts of Europe any of these methods may be seen, or others. In France, for instance, the SSADM-like *Merise* is in use.

2.8   PG TOPIC: Which methodology?

The question of which IS methodology to use in particular circumstances is of great importance. Jayaratna (1994), among others, has considered the philosophy behind various development methods, and which might be used for what types of system. His NIMSAD system attempts to pick horses for courses by offering a scheme for selecting the correct methodology for the particular circumstances of a project in question.

---

[4]   *Methodology:* In general, a coherent set of methods used in carrying out some complex activity. The word is most frequently used in terms such as *programming methodology* or *(system) design methodology*. In the UK, the word *method* is usually preferred for this meaning. (DOC)

2.9   Tutorial 2.1

As an IT manager, what would you say about hiring the following sorts of people as systems analysts?

- a whiz programmer who has trouble with interpersonal relationships;
- someone who has never been a programmer but is the life and soul of any party;
- a known gambler who is very good at calculating the odds;
- a nervous person who is a very good listener;
- a rather brash person who can talk the hind leg off a donkey;
- someone who is very methodical and rather dull.

Requirements analysis: the discovery of what is needed

3.1 The purpose of requirements analysis

As we have said, a major job of the systems analyst is to find out what is required of a new system.

This activity has also been called:

- requirements analysis;
- requirements acquisition;
- requirements determination;
- requirements elicitation;
- requirements engineering;
- and a number of other things.

In a much more detailed account, Donal Flynn (1998) suggests that 'requirements determination' is the best term:

> The **requirements determination** phase consists of several stages: **problem definition, feasibility study, requirements acquisition** and **requirements modelling**. The aim is to obtain a description of the user requirements which are expressed in terms of user concepts.
>
> (Flynn, 1998, p. 101)

However, we are using the fairly common term *requirements analysis* (RA) here. The aim of requirements analysis is to gain a thorough understanding of the requirements of the new system, and to firmly establish the direction and viability of the rest of the project.

It sets up the analysis phase of the project by creating plans and agreeing the initial scope of the study.

> *The purpose of RA is to:*
>
> - identify what data and what processes are needed in the new system;
> - determine the functional and non-functional requirements of the new system;
> - set out several business system options in language that managers will understand;
> - specify the requirements without expressing computer alternatives and technology details.

Hint

The difference between what users **want** and what they **need** is interesting. The user may want lots of things, including the sexiest system on the planet. However, a negotiation must take place to determine what is really needed.

3.2   System investigation

For many proposed information systems (ISs), the aim is to computerize the current paper-based IS. During requirements analysis, one of the more important activities is to investigate this current system. What is needed is a thorough understanding of what is happening in the IS now in use.

This is meant to enable the systems analyst to clarify what the users need and want out of the new system. RA is also intended to define and prioritize the users' requirements, and to consider security, control and audit requirements.

The outcome of the investigation is the ability to model (using diagrams and words) the current IS. The present system is working in physical form, so it is called the *Current Physical System*.

3.2.1   Interviews

Perhaps the main way of discovering what the users want and need from their new system is to ask them during an *interview*. Interviewing is also one of the most important tools to investigate the workings of the current system. During these interviews the analyst will be trying to find out how the current system works, the problems that the users perceive in the current system, and what the users want from a new system.

One of the difficulties with interviews is finding the right members of staff to talk to. The people interviewed should include those who are on the ground floor actually working within the system, and those who will be using the new system. Managers will also need to be spoken to.

The quandary with speaking to managers is that they often believe they understand the way the current system works, but sometimes they do not. Even those managers who used to work with the current system may now be under many misapprehensions about what happens nowadays. Their ideas may be out of date, and they may have been misinformed about what happens by the current staff.

However, it may well be that the analyst is only able to interview managers. This is especially the case for a brand new system for which the ground floor staff have not yet been hired. In such cases, there will probably only be managers to speak to.

3.2.2   Good interview etiquette

There are a few words of advice that are worth considering before going to interview someone.

Never go empty handed:

- Always do your homework beforehand;
- Know something about the person and department you are going to see.

Assume very little:

- You may not have understood what you think you know;
- Someone else may have misinformed you.

Do not tell the interviewee what you should be being told:

- It is the interviewee who knows, not you;
- Your task is to find out what the interviewee knows;
- Telling them what you think they know may result in the truth never being told.

Try asking 'What happens in your department?'

- This simple question should elicit most of what you need to find out.

Do not interrupt:

- Once the interviewee is talking, any interruption may result in both of you missing out some details.
- If you *must* interrupt, it is your professional duty to know where to restart again.

There are short lists of do's and don'ts that are also worth keeping in mind.

DO NOT ...

Be superior:

- There is little more irritating to someone in work to be spoken to by someone who obviously thinks that they are more important than you.

Be arrogant:

- Body language or a way of speaking can give the impression of arrogance. It is not a way to encourage people to talk to you.

Show that you know how things should happen:

- This can diminish the interviewee's perception of the importance of their own job and tend to make them less informative.

Show that you think the interviewee's job is:

- Stupid.
- Incompetently done.
- Boring.
- Useless.

Show that you think they are lying to you:

- There are ways of politely checking when something they tell you does not agree with what you think you know already.
- You must be able to handle such situations sensibly and without giving offence.

On a more positive note:

DO ...

Introduce yourself properly:

- Do not walk in, sit down, and begin asking questions.
- Say who you are and why you are there.

Ask if you can take notes:

- No one will expect you to memorize what is going to be said, but it is courteous to ask.
- Certainly **never** make an audio recording of the proceedings without asking permission. Once that is discovered, no one will speak to you ever again.

Be nice:

- Nasty analysts are inefficient and misinformed.

Be humble:

- You are asking them for what they know.
- Do not act superior or arrogant.

Ask for clarification if you need it:

- You will need to be sure about what has been said. As mentioned before, clarification is better left to the end of the interviewee's remarks.

Recap at the end:

- It is a good idea to tell the interviewee what you think you have found out. This may well result in mistakes on both sides being corrected. It also may open up areas that were missed initially.

Be thankful:

- At the end, offer your gratitude for their time. They will then be more likely to talk to you again.

There are additional ways of digging out information.

3.2.3   The paper chase

This is very important. The analyst will need to collect all the different types of paper that allow the current system to run. There will be forms, memos, notes, and reports. The analyst has to find every piece of data that is present in the current system, and also know where it comes from and goes to.

3.2.4   Observation

The analyst may need actually to watch what happens. There might be an occasion to sit with a member of staff to see what happens in detail, or even to follow people about as they do their work within the current system.

In some cases, the analyst might need to set up a small, dummy version of the IS so as to watch the staff carefully as they carry out their various tasks.

3.2.5   Problems and requirements

During the investigation of the current physical system, many problems will be uncovered, as will several functions (requirements) that the users do not have but they would like to have.

Problems and requirements should be noted in a list, often called a *problems/ requirements list* (PRL). This is a form that has a brief description of the problem or requirement together with a suggestion of possible solutions to solving the problem or meeting the requirement. Ultimately, the entries on the PRL must be entered into a formal document known as the *Requirements Catalogue*.

3.3   Requirements catalogue

Each problem and requirement is further refined and prioritized so as to become an entry in the document called the requirements catalogue (RC).

So, this important document is prepared during the analysis stage of the systems development life cycle. The content of this document can be referred to as the *expression of requirements* for the system.

A particular requirement has to be carefully documented so that, throughout the development of the new system, it can be used as a benchmark of what the new system is supposed to do. This validation should happen at all levels of development. For instance, a computer program that does not refer to any requirement in the RC, may be nothing more than a program written only for the benefit of the programmer.[5]

> *Requirements should, as far as possible:*
>
> - be quantifiable;
> - not be duplicated elsewhere;
> - be detailed enough to base decisions on them.

To be quantifiable, or quantified, some numbers (quantities) must be involved. It is not good enough to say that the new computer system must be up and running 'most of the time'; the requirements definition should state that the system should be available between certain times of the day and should stay operating for some average amount of time before crashing. To be very quantifiable and statistical, a useful concept is the *mean time to failure*. Such a requirement could be that the system is needed to be available for 60 days, on average, before it fails.

Similarly, it is not enough for a requirement to be that the computer will respond to a user 'as quickly as possible'. Response times should be quantified in seconds for particular functions. So, the requirements definition for a simple enquiry might be that the response time should be no greater than 3 seconds.

---

5   Such a program can be referred to as a *horse-racing program:* a program written so that the programmer can place better wagers on the horses.

Expression of requirements: A statement of the requirements that some envisaged computer system ... is expected to meet ... A good expression of requirements should be one of the earliest products of any system-development project, and for a project of significant size it is of crucial importance, not least because errors at the requirements stage tend to be the most expensive to correct.

DOC

A more complicated function might be expected to take no more than 10 seconds, and so on.

Zero response times are, as a matter of fact, impossible, but they can seem like that to a user. In the requirements definition, the mention of 'at most 1 second', may be more politic than promising an 'immediate response'.

The requirements catalogue contains details of:

- what is wrong with the current system;
- desirable features for the proposed system;
- the system's **functional requirements** – what the system has to do;
- the system's **non-functional requirements** – such as the level of performance, and resource usage.

It should be noted that not all entries in the RC may be incorporated into the new system. As mentioned above, some apparent *needs* may merely be *wants*. All requirements should be examined carefully to check that it is a real need for the new system.

The mean time to failure of a system is the average (mean) time that the system is expected to be available before it fails.

3.4   Functional requirements and non-functional requirements

We have mentioned these terms a couple of times, and we need to be clear about them. Requirements come in two types: a requirement is said to be *functional* or *non-functional*. The area between the two is a little grey, but the distinction is commonly used, so you need to understand it.

A *functional requirement* dictates **what** the system should do. What facilities are required and what activities the system should carry out.

Response time is the time between the user entering the data (often by pressing an enter key) and the screen displaying the computer's reply to that data.

Functional requirements can concern:

- descriptions of required functions;
- outlines of reports—both hard and soft output;
- online queries and updates;
- data storage, retrieval and transfer.

Non-functional requirements (NFRs) address other needs. These are often limitations or constraints on the way the system carries out its functions. Flynn (1998, p. 148) notes that NFRs have been 'traditionally viewed as global properties of a system', and, in the past, have only been informally addressed.

Aside

You might think that it is always a good idea to make a response time as short (as fast) as possible, but consider the way a user might think. Say the user enters what seems to be a lot of data or asks what they think is a complicated question.

If the computer responds immediately, the user might doubt that serious consideration has been given to that difficult task. Unless some obvious sign is given that the data has been accepted properly, or that this is the real answer to the enquiry, some users might doubt the response and be tempted to try again.

Far be it for us to suggest that, in some very special cases, the response time might be artificially lengthened (the technical term is 'slugged'[6]) in order to persuade some users that the computer has actually done something.

Examples of non-functional requirements might be:

- response times;
- mean time to failure;
- security needs;
- access for disabled users.

Each requirement in the RC should include some prioritization—an idea of how important the requirement is to the system or the users. It may be **mandatory**, **desirable**, or **optional**.

Also, there should be no harm, even at this early stage, of giving some idea of how the particular requirement might be met.

Figure 3.1 gives an example of an entry in a RC.

3.4.1　Requirements specification

The requirements specification is designed to document the analyst's understanding of exactly **what** must be done to solve the problems. It contains the definitive statement of the requirements for the new system, and defines exactly **what** the system must do in as quantifiable a way as possible. An example would be that the response time following inputting some data must be less than 5 seconds.

The specification of the requirements defines the user's needs clearly. It is designed to state **what** the new system should do rather than **how** it is to be done, and incorporates a lot of cross checking between, for instance, the DFDs and the LDS. There has to be a high level of customer and user involvement to check and confirm the requirements.

The document produced as the deliverable of the requirements specification is a customer or user-oriented document using business terminology as far as possible. Users can (and probably should) sign for acceptance of the required system specification.

3.5　Example of an entry in the Requirements Catalogue

The example in Figure 3.1 is adapted from Weaver *et al.* (1998, p. 30)

---

[6]　　A dictionary definition of *to slug* (in this sense) was not easy to find. However, the Shorter OED has from the seventeenth century: '**slug**: *v.t. Relax, slacken, make sluggish*' – which seems to fit.

Requirements catalogue entry			
System: *SWR*			
Source: *Goods in manager*	**Priority:** *D*	**Owner:** *F. Bloggs*	**Requirement:** *Id. 14*
Funtional/Non-functional requirement: *To provide details of overdue deliveries*			
Description: *Availability* *Access*	**Target value:** *On-line 9.00–18.00 hrs Mon–Sat* *Goods in manager*	**Acceptable range:** *Response within 10 minutes*	**Comments:** *Early morning availability essential to clarify delivery clashes*
Benefits: *Enable monitoring of supplier response and help in chasing up overdue deliveries.*			
Comments/Suggested solutions: 			
Related documents: *Interview notes No. 3, Function 14*			
Related requirements: *No. 15; No. 16 (non-functional)*			
Resolution: *Part of all BSOs 1; accepted and followed through*			

Figure 3.1   A requirements catalogue entry.

3.6   PG TOPIC: The impossibility of complete requirements analysis?

One of us presented a paper some years ago (Chester, 1995) that asked several philosophical questions about the elicitation of requirements. The argument of the paper is that the terminology utilized by users may not lend itself to the formal, fixed, tightly bounded world of computing.

User terminology is important because the computerized system will be the encapsulation of what the users mean by what they say. So it is very important to understand it. Furthermore, the tendency is for IS personnel to need to classify this terminology in a rather rigid way. This is especially true of object-oriented approaches that particularly deal with classification and inheritance.

The philosophical problem comes from the work of Ludwig Wittgenstein. In his *Philosophical Investigations* he suggested that even in ordinary language it may be impossible to classify words in a straightforward manner. The example he uses is the word *game*. He notes that there are many terms that would be classified as games, but he writes that nothing can be found that links them all together in a logical way. Although poker is a game and professional football is a game, so is pat-a-cake and hopscotch. He denies that there is any underlying attribute or group of attributes of these activities that could be inspected to show that they all fit into the class of game. What actually happens is that people simply agree to say that an activity is a game, and it becomes common usage.

Instead of some underlying commonality, he offers the concept of 'word families' and suggests that the terms that are classified as games can be put in a continuum where each word has things in common with its neighbour to make them both games. However, these common factors alter as we move along the continuum, so that, by the time we get to the end, there is nothing in common between the first and the last. However, because of their family resemblance, we call them both games.

The application of these ideas to ISD suggests that when a user says that several items are all called aeroplanes, there may be no shared logical feature to enable us to classify them safely. Furthermore, when presented with some new item, it may be impossible to ensure that it is an aeroplane or is not.

The Chester (1995) paper should be read to appreciate more of the problems and (failed) attempted solutions to this. The outcome is that a firm, logical, clear-cut classification of the terms we need to understand in our systems analysis may be impossible.

3.7 Tutorial 3.1

Think about some of the systems with which you have been associated and involved. Set down several problems that you feel these systems have, and offer some solutions to those problems.

Some possible systems are:

- the student loans system;
- the college application system;
- the college or public transport system;
- the college meals system.

In your solutions, be as bold as you can. For instance, a solution to most of the problems with the student loans system is to do away with students entirely. Not that we would advocate that for a minute.

4 Report writing and presentations for the systems analyst

4.1 The systems analyst speaks

We suggested earlier that the job of a systems analyst is not a solitary occupation. The modern organization requires that ideas are promulgated, sold, and criticised. Systems analysts will have many opportunities to disseminate information to their colleagues in IT or the general business community, be they managers or staff.

This chapter deals with the important topics of writing formal business reports and giving formal presentations to staff and managers.

4.2   Writing formal reports

Reports come in all shapes and sizes. Here we shall be discussing the type of report that would be written about the findings of a systems analysis exercise, and written for the management of a company who, we shall assume, have little knowledge about computers and computerization. We shall be concentrating on the structure of such reports and the language that is used in their production.

The aim of such reports is to get across vital information to those who need to know it; and to get it across in the most clear and straightforward way so as not to upset or irritate those who have to read it.

4.3   What formal reports are not

Sometimes it can be useful to see what should not be done in order to understand what ought to be done. In this case, those who are not familiar with producing formal reports—or worse, those who have seen or have already produced bad formal reports—may need to be disabused about what reports are not (see box left).

4.4   Structuring reports

It is better to follow a standard layout for a report than to produce it in some haphazard fashion. We suggest that the structure of all formal reports follows a pattern such as:

- Title page;
- contents;
- summary;
- introduction;
- findings or main body;
- conclusions and recommendations;
- references and bibliography;
- appendices;
- other sections.

4.4.1   Title page

This is the first thing that a reader sees of your report. It should be a complete page, containing the full title of the report along with any subtitle. The author(s) name(s) should also appear as well as the company's name and logo.

4.4.2 Contents

This is also known as the *contents page* (although, in a large work like this book, it may run to several). It lists all the separate sections or chapters of the report, giving their titles or chapter headings. It might also list sub-headings for subsections of chapters. The contents of this book should give a good idea of what would be expected of the contents section of a report.

4.4.3 Summary

The summary of the report is sometimes known, particularly in a paper for an academic journal, as the *abstract*. As the name suggests, it summarizes—**briefly**—the contents of the report and its main findings. It contains the major findings of the report and the most important conclusions and recommendations. Its usefulness is twofold:

- A reader will look at it to decide whether or not there is any point in reading the rest of the report.
- A reader will look at the summary to be given the flavour of the full report, so they know what to expect in the entire document.

Pitfall

Often we read student summaries that read like introductions to the report. There is an introduction to come; the summary is a short and concise crystallization of the whole report.

4.4.4 Introduction

This section sets the scene for the entire report. It may give details about why the report has been written, and give some historical information about the company or department or system with which the report is concerned.

Subsections of the introduction could include:

- *History* of the report: why and how it has come to be written.
- *Terms of reference*: the set objectives of producing the report.
- *Guide to chapters* or sections: in a large report, it might be useful to the reader to be given some foreknowledge of what to expect in each section. This may allow the busy reader to read more selectively. Such a chapter guide is to be found at the beginning of this book.

4.4.5 Findings

This is the main body of the report. It may well need to be separated into several sections or chapters, each dealing with the different points that need to be described.

Detailed figures and listings (such as statistics or program code), that would clog up the flow of the text and distract the reader, should be put into the appendices at the end of the report. The text in the findings would refer to the appendices, should the reader wish to consult them.

4.4.6 Conclusions and recommendations

This section may be split into a section or chapter on conclusions and another on recommendations. Sometimes the report is the sort of document that does not need to contain recommendations, but, generally, some sort of conclusions will be called for.

Neither of these parts of the report should introduce brand new information. The new information will already have been covered in the Findings chapters. The conclusions is the place where the findings are used to support the conclusions that the writer of the report has deduced from what has been discovered. Generally, no new references or citations should appear here, unless they are being used to support the way the conclusions or recommendations are being construed.

4.4.7   References and bibliography

These are two different types of section and may well appear as two different chapters. The references section contains a list of those works for which an actual quotation (in reported or direct speech) has been given in the preceding report, or from which a diagram has been taken.

The bibliography contains a list of those works that the writer has consulted, from which no actual quotation has been used. These works often give background information that the reader might wish to consult to take their reading further.

4.4.8   Appendices

This is the place for all those statistics, interview notes, and program listings that you think some readers would need to look at but would have disturbed the flow of the text in the findings section. Several appendix sections are possible. Traditionally, an appendix section name is based on a letter such as Appendix A or Appendix B.1.

4.4.9   Other sections

Index

There may be a need for additional parts to a report. In a large document, such as a book or a government report, there may be a need for an index. The index at the back of this book should give a good idea of how to produce one of these.

Distribution

Some reports need a distribution section. This could be at the front or the back of the document; it has the people and their addresses to whom the report is to be sent.

4.5   Section numbering

In order to make each part of a report distinct, and to allow a reader to point to one of these parts without confusing it with some other part, it is strongly recommended that chapters (or sections) and subsections of formal reports are numbered. In certain cases, even individual paragraphs may be numbered as well. The numbering system used in this book should act as a good guide to what is required.

As in this book, chapters are given an integer such as Chapter 1 and Chapter 10. Subsections of a chapter are given two integers separated by a dot, for instance, section 1.1 and section 10.2.

If necessary, parts of subsections may need to be numbered at the third level, such as 1.1.1 and 10.2.4. You will not miss the correspondence with the numbering system used in Chapter 9 later.

4.6 Citations and references

Any direct reference to someone else's work should be acknowledged by making it clear who produced that work originally. Failing to credit other people where you ought to, or presenting their ideas or words as your own, is at least a moral failing and at worst could lead to a charge of *plagiarism*.[7]

A reference to someone else's work is called a citation—you are citing their publication. Citations may be of books, journal articles or papers, web pages, trade press reports, and other people's formal reports. The reference made could be a direct quotation, within quotation marks, thus: **Chester and Athwall (2002) state that 'A reference made to someone else's work is called a *citation*'.** This could also be shown as an indented set of words often separated by blank lines from the rest of the text and italicized, thus:

> *A reference made to someone else's work is called a citation.*
> (Chester and Athwall, 2002)

These examples are in direct speech and quote the exact words used by the other author. Indirect speech can also be used, thus: **Chester and Athwall (2002) wrote that plagiarism is at least a moral failing.** This is the equivalent of the direct quotation: **Chester and Athwall (2002) wrote, 'plagiarism is at least a moral failing'.** There are two main ways of showing whose words or ideas you are using. These are the Harvard system and the Vancouver system.

The **Harvard system** is in use in this book and is shown by the examples above where the words of Chester and Athwall from a work published in 2002 are introduced by the authors' names followed by the year of publication in parentheses or round brackets, thus: **Chester and Athwall (2002).**

We feel that a citation reads more naturally if it is made a part of an ordinary sentence, as shown in most of the examples seen so far. In these cases, the name of the author or authors is included as part of the ordinary sentence and the year

---

[7] *Plagiarize:* to steal from (the writings or ideas of another). (*Chambers CD*)

Pitfall

Putting the author(s) in parentheses after a citation should not be used to refer to a large section of text where some or all of the ideas come from some other work, but where the reader will not be able to tell what words or ideas are from the cited author(s) and which belong to the work now being read. In students' reports it is common to find a whole paragraph, or even several paragraphs, with an author citation tacked onto the end, with no clues as to whether the one preceding sentence, or the whole paragraph, or some unknown number of preceding paragraphs comes from the cited work or not.

of publication follows the name(s) in parentheses. Often (especially for books) the page numbers should be shown as well, thus: **Moreton and Chester (1997, pp. 64–65) write about developing a framework for strategic planning of IS/IT**.

In some quotations, as in the dictionary citation in the note and in large quotations, it is better to show the names and date immediately after the quote, enclosing them all in parentheses, thus: (Chester and Athwall, 2002, p. 108).

In the Harvard system, the works cited are listed in the references section in order of the authors' names, as seen at the end of this book.

The **Vancouver system** of citations replaces the information included in parentheses with a superscript number that refers to the list of works in the references section. So, using the previous examples, this other system would have:

Chester and Athwall[1] state that 'A reference made to someone else's work is called a citation'.

A reference made to someone else's work is called a citation.[2]

Moreton and Chester[3] write about developing a framework for strategic planning of IS/IT.

Often, this number is not put as a superscript but in square brackets. Thus:

Chester and Athwall [1] state that 'A reference made to someone else's work is called a *citation*'.

A reference made to someone else's work is called a citation [2]

Moreton and Chester [3] write about developing a framework for strategic planning of IS/IT.

The references at the end are then listed in numerical order:

1. Chester, M.F. and Athwall, A.K., *Basic information systems analysis and design.* McGraw-Hill, Maidenhead, England, 2002.
2. Chambers CD.
3. Moreton, R. and Chester, M., *Transforming the business.* McGraw-Hill, Maidenhead, England, 1997.

Some places prefer Vancouver and others Harvard. There are arguments for each. We can say however that it is no small task to switch from one to another once a work is complete.

4.7 Standard of English

It may seem obvious that good quality English should be used in the writing of any report, yet we often come across students who believe that the level of their language skills is unimportant. For any formal report, good English is essential. There is little more irritating to an educated reader than being presented with a report that uses careless or slovenly language. This includes correct spelling as well as grammar. There are many books to help in this endeavour, and we suggest the reader looks at one or more of these. However, the very best advice that we can give for those whose English is poor, is to read more. The reading of good modern English, as found in broadsheet newspapers and books by established authors written in modern English, is by far the best way to acquire the habit of good English writing.

Modern word processors offer some aids to the aspiring writer such as spellcheckers and grammar checkers. They can be useful to all of us. However, the main problem with spellcheckers is that they will not stop you using the wrong word as long as you spell it correctly. Also, grammar checkers are useless if you do not understand whether what they are suggesting is better than the words you originally used. Nevertheless, always run the spellchecker just before printing the report. Murphy's law for writers means that it is that small change you made just before printing that will contain the spelling mistake that blights your entire document.

The style of English adopted for a report is important too. Formal reports should use formal English. This means that the English used there is not the same as you would use in everyday speech—spoken English is different from formal written English. As a rule, do not use contractions such as **can't**, **won't** and **shouldn't**. Use **can not** or **cannot**, **will not**, and **should not** instead. Also, tend to write in an impersonal way. This means not using the words **I** (the most commonly used word in English) and **we** so often. So, for example, instead of writing **I did this or that**, write **this or that was done**.

Bad examples

- What I think's that the way they do things in this place is real bad, & can't see how they get anything, why can't they do it right I say. Innit?
- The comprehensibly obvious outcome of my deliberations is that this party has an incontrovertible and incontestable incapability.

Some writers think that a report is an opportunity to show how clever they are by using words that they do not expect their readers to understand. In reading a novel or even a basic textbook the reader might be expected, once in a while to have to turn to a dictionary. However, in a business report this should rarely be the case. Ordinary, simple, educated English is sufficient. There is also the greater danger that, if you are using words that your readers do not understand, you will also be using words that *you* do not understand. We often read reports that show that the level of English being attempted is far too high for the ability of the person who is writing it.

After many years of marking the writing of students, we have encountered a large assortment of common errors made by them. Many of these errors involve the inability to use apostrophes properly, but there are other frequently found mistakes too. We include here some of the most common errors we have noticed, in the hope that it will improve your English when you need to write that important report.

4.8 Notes on English

The following may prove to be useful to those who are not used to writing essays and reports. There is, however, no substitute for reading good English.

A few notes on English in reports

After reading and marking many essays and reports, we have noted a few problems that occur frequently. So we thought it would be worthwhile to hand out some advice on these matters.

1   Apostrophes

(The mark used in words like '**Myrvin's**' in '**Myrvin's notes**')

1.1   The 'grocer's apostrophe'

This is a rather flippant term for the use of the apostrophe in error before almost any **s** at the end of a word, particularly in a plural.

E.g. 'Sweet Plum's', 'Potato's—home grown'

For some reason, grocers, in shops and on market stalls, are famously fond of this mistake. Do not incorporate it into your reports.

There is also no reason to use apostrophes in the plurals of abbreviations. Do not use '1990**'s**' or '**PC's**' where you are referring to the plural. The modern way is to write '**1990s**' and '**PCs**' (see also, 1.2.1).

1.2   Possessives

These are words that denote ownership etc., and are usually created by adding an apostrophe (') and an **s** to the noun.

E.g. 'Myrvin's notes' = 'The notes belonging to Myrvin'.

This is a valid use of the apostrophe, but there are a few things to watch out for.

1.2.1 Plural possessives

If the word is a plural ending in **s**, then the apostrophe **follows** the **s**.

E.g. 'The notes of the two lecturers' can be written
'The **lecturers'** notes'.

If the mark came before the **s**, it would suggest only **one** lecturer was involved.

So: 'The **lecturer's** notes' = 'The notes of the lecturer'.

Using the plural form for abbreviations outlined in 1.1 allows you to write:

'The **1990s'** fashions are crazy', or 'Those **PCs'** keyboards are faulty'.

1.2.2 Words ending in s

You may see the following: 'I climbed St **Giles'** church tower', where 'St Giles' is one church not more. There is some confusion about how to handle such words. Some say you should write it as above, but pronounce it as 'St **Giles's** church tower', but others would not pronounce the extra **s**. Some say you should pronounce it *and* write it as 'St **Giles's** church tower'.

1.2.3 Odd possessives

There are a few ownership words that you might think should have an apostrophe but do not. The main ones are '**its**': belonging to 'it', as in 'The company and **its** employees'; and '**hers**': belonging to her, as in 'This coat is **hers**'.

They do not have apostrophes. You might remember that the possessive '**its**' is spelled like that to distinguish it from 'it is' (see below). However, there seems no similar justification for the word '**hers**'—just remember it. Other examples are: '**yours**', '**ours**', and '**theirs**'. However, the possessive from 'one' is still '**one's**'.

E.g. '**One's** husband is the Duke of Edinburgh; **yours** isn't, and **theirs** aren't either'.

1.3 Missing letters

The apostrophe is also used to mark the place where a letter or letters have been missed out. For instance the word '**it's**' is the same as 'it is'. The second **i** is missed out, and the apostrophe denotes where it should be. Such words are called contractions.

E.g. '**It's** the English that matters'.

Other examples:
'Avtar**'s** here' is the same as 'Avtar is here'.

'They're over there' means 'They are over there'.
'Don't do that' equals 'Do not do that'—the second o is missed out.

In formal essays and reports do not use such contractions—write the words in full.

2. 'There' and 'Their'

These are often confused.
The word 'their' is a possessive, meaning 'owned by' or 'to do with' **them**.

E.g. '**Their** essays are over **there**.'

Make sure you are using the correct word. We have even seen 'they're' in mistake for 'their'.

3. 'i.e.' and 'e.g.'

Also, often confused.
The abbreviation '**i.e.**' is short for '*id est*' which means 'that is'. It introduces the **complete** list of examples or the **only** example. It can be used as in the following:

'Myrvin had three cats, **i.e.** Freud, Charlie and Hardy, and one dog, **i.e.** Rover'.

The abbreviation '**e.g.**' is short for the Latin *exempli gratia* and means 'for example'. It introduces selections from a larger list, as in:

'Some of Myrvin's cats, **e.g.** Freud and Charlie, were almost completely black'.

Be careful to put the full stops in the correct place. However, in formal reports and essays it would be better not to use these abbreviations. Replace **e.g.** with phrases like 'such as', 'for example', or 'for instance'; and **i.e.** with 'that is', or by rewriting the sentence so as not to need it.

By the way, do not put 'etc.' at the end of a list that has been introduced by 'e.g.' or 'such as'. This is because the introductory part is already telling the reader that there are more items not being mentioned.

4   Combining words wrongly

4.1 'alot'

Students may produce a new word that does not belong in English. The main example we have seen recently is the pseudo-word 'alot'. There is no such word as '**alot**'; the phrase is '**a lot**', as in 'We make **a lot** of mistakes'. Allowing the spellchecker to replace it with 'allot' is even worse. Do not invent false words by combining two words into one.

4.2 'apart'

This word, while alright in itself, is often used where the writer means 'a part'. So 'This section is **apart** of the English notes' is a mistake for 'This section is **a part** of the English notes' and means almost the opposite of what was intended.

5. Affect and effect

Once more, these are often confused.

'**Affect**' will generally be a verb. To **affect** something is to influence it. '**Effect**' will generally be a noun. The **effect** of something is what happens as a result of that something.

> E.g. 'The **effect** of these notes should be to **affect** your English for the better.'

There is a verb 'to **effect**', which means to produce something or to bring something about, and there is also a noun 'the **affect**', which refers to the emotions. Use these two words very carefully.

6. Maybe and may be

The first is an adverb meaning 'possibly' or 'perhaps'. It is unlikely to be used correctly in an essay.

> So, '**Maybe** this is correct usage, and **maybe** it isn't' is correct; but 'This **maybe** a cat' is not.

Mostly, the correct usage is '**may be**'.

> E.g. 'This **may be** the case or it **may** not **be** the case.'

If you find yourself writing '**maybe**', you most probably mean, '**may be**'.

7. Use of articles

The word '**the**' is the definite article, and the words '**a**' and '**an**' are the indefinite articles. The first is used to introduce a noun that refers to a specific, single example. The indefinite articles refer to any example. Their use is complicated, and difficult for those whose first language does not use them; but here are a few points.

> E.g. '**The** cat sat on **the** mat' refers to a particular cat and a particular mat.
> '**A** dog sat on **an** ant' refers to any dog and any ant.

Most sentences need ordinary nouns to be introduced by one or the other article. However, proper nouns (names of people, organizations etc., spelt with a capital letter) do not need articles.

> E.g. 'Myrvin is typing this'; not 'The Myrvin is typing this'.
> 'Microsoft is a large company'; not 'A Microsoft is a large company'.

8.　IS and IT

In reports and essays about computing, these abbreviations are very common, and commonly used wrongly. We often see 'I.S' and 'I.T', with the final full stop missed. The correct forms are 'IS', 'I.S.', 'IT', and 'I.T.'. Choose whether you intend to put full stops or not, and stick to it. We have also seen 'V.A.T' instead of 'V.A.T.' or 'VAT'.

9.　Quotation marks and slashes

The use of **quotation marks** seems very popular with students. Some reports and essays are liberally sprinkled with double and single quotation marks, most of which are completely unnecessary. Often they are used to surround a word or phrase that the student feels is not quite formal English; or it may even be felt to be slang. If this is the case then our advice is simple: do not use that word or phrase. Sometimes, the student puts the quotation marks around a word or phrase that is felt to be slang, but in fact it is not. In such a case the marks should be omitted. Either way, the use of quotation marks should be used with extreme care, and probably avoided completely.

　　Slashes are sometimes very popular too. Instead of using the formal linking words such as **and** and **or**, people like to put a /. In the hands of these people, a sentence modified from the previous paragraph could be written:

Sometimes / always, the student / lecturer puts the quotes / quotation marks around a word / phrase that is felt to be slang / informal.

This is lazy and tiresome to read. We feel that the only (barely) acceptable use of such a device is in **and/or**, and, less acceptably, **his/her**. If you want to say **student** *and* **lecturer** or **student** *or* **lecturer**, then write the one you mean, and not **student/lecturer**.

10.　Numbers

There are many references to numbers in reports, and students often wonder whether to use the numerals themselves or to spell out the number in letters. It looks very odd to see a sentence like:

　There are **2** reasons and **one hundred and twenty-three** possibilities.

A rule of thumb that works well is to use the letters for numbers under, say, ten (10), and numerals for larger numbers. So, the previous example would be better written as:

　　　There are **two** reasons and **123** possibilities.

11.　Other mistakes

The words 'lose' and 'losing' are often mis-written as 'loose' and 'loosing'. The first two words are to do with a loss, while the second concerns something being set loose or set free.

> The correct phrase is 'with regard to' or 'as regards', not 'in (or with) **regards** to'.
>
> In UK English, 'practise' is the verb, and 'practice' is the noun; and 'license' is the verb and 'licence' is the noun.
>
> E.g. Solicitors **practise** in their **practices** to **license** our **licences**.
>
> If it helps, recall that 'advise' (with an s) is the verb and 'advice' (with a c) is the noun. US English does not use 'practise' or 'licence' at all—but then they do spell 'defence' as 'defense'.
>
> The past tense of 'input' is not 'inputted'; it is '**input**'.
>
> The past tense of the verb to lead is spelt '**led**', not 'lead'.

4.9 Presentations

Any competent systems analyst may find that they are required to present information to a group of people. This activity is a *presentation*; and we feel it important, in a book incorporating the basic practicalities of the work of a systems analyst, to write something on this topic. Again, by way of introduction, it may be useful to give a list of what a presentation is not (see box left).

4.10 Initial considerations

Some important questions should be asked before embarking on the plan for any presentations.

- Who is the audience?
- Why are you here?
- Why is the audience here?
- What will be the content of the presentation?
- How are you going to deliver what it is you want to say?
- What structure will it take?
- What non-verbal presentation aids should be used?

4.11 Structure of the presentation

All talks need a good framework so as to convey the message logically. The introduction is particularly important, setting, as it does, the tone for the whole presentation. The conclusion is also significant, allowing the opportunity to round off the talk and leave a good impression with the listeners.

In a group presentation, where several people participate, it is good practice to announce in the introduction who is speaking about what. Also, as each presenter begins their part of the talk they should introduce themselves and say what they are going to cover. At the end of their piece they should say they have finished and introduce the next speaker and the next topic. This allows for a smooth transition from speaker to speaker. A group presentation should be well balanced, so that each presenter has about the same amount of time as any other.

4.12 Non-verbal behaviour

Pitfall

Irritating habits:
- playing with your ears (and other parts of the face);
- twiddling your thumbs;
- clicking your pen;
- clicking your tongue;
- shifting from foot to foot (and other dances).

A good speaker appears confident (even when they are not), looks at the audience (even when they are not pretty), and avoids irritating and distracting habits (see box left).

Your presentation should be professional, which means you should dress and act appropriately. People giving talks (or even attending meetings) who look as if they have just come in from an encounter underneath an oily car parked in a gutter are an embarrassment to all concerned—unless of course they are speaking to a meeting of other garage mechanics at work.

4.13 Verbal behaviour

The language you use for the talk might also need some attention. Personally, we do not object to regional accents or accents from overseas. Nevertheless, you must be understood by your audience, so you may need to tone down your accent if it is particularly strong. In addition, you may wish to change the way you express yourself, especially if your local dialect is not going to be readily understood by your audience. One of us attended an academic conference where a speaker thought it was clever or attention grabbing to sprinkle her talk with swear words. Perhaps she wanted to be thought of as an ordinary person rather than a boring academic. This smacks of desperation and may upset your audience rather than gain their sympathy.

4.14 Use of visual aids

Visual aids are generally expected in a modern presentation. Someone simply standing at the front and talking is not really sexy enough for today's audiences. Visual aids such as overhead projector (OHP) slides are considered to be the minimum required. There are also slide projectors, computer projectors, flip charts, and video displays. Worth keeping in mind as well are the rather unsexy aids of black and whiteboards.

If you are using such aids it is important that the venue is checked out beforehand to ensure that they can cope with them. Most will have an OHP and slide projectors, but computer projectors, and, maybe, flip charts are certainly worth enquiring about. Many venues may also have difficulties in providing blackboards and whiteboards.

Each piece of equipment needs to be thoroughly checked before the presentation. There is little more irritating to an audience than to be kept waiting right at the beginning of a talk because the OHP fails to work when it is first switched on. The machine should also be adjusted before the talk so that the slide is projected properly onto the screen and is in focus.

The creation and use of OHP slides takes a little training. Each slide should not be crammed with information and the lettering must be large enough to be seen at the back of the room: size 20 font is probably the minimum. Generally, it is better not to expose the whole slide at once. Members of the audience will try to read the whole slide as soon as they see it. This means that they will be looking at the bottom of the slide while you are talking about the top, and they will therefore not be concentrating on the point that you wish them to. There are three main ways of trying to ensure that they do look where you want them to. You could put a pointer (a pen or something) on the slide that indicates the line you are talking about. A pointer could also be used to point to the screen. This could be a long stick or a laser pointer. However, in both these cases they will still be reading the whole slide. Preferable to these would be to cover the parts of the slide that you have not yet come to. A piece of paper or card will suffice. Computer projector presentation packages can be set up to put each line on the screen as you need it.

4.15 Preparation and practice

It will be necessary to rehearse your presentation—as a group if it is a group talk. You need to ensure that the talk takes the correct length of time and runs as smoothly as possible. You will also need to practise with the visual aids you are going to use.

4.16 Assessment guide for group presentations

The following gives the assessment guidelines and mark sheet that we have been using for some time to mark student presentations. We hope they will help you in giving professional presentations of your own.

University of Wolverhampton School of Computing and Information Technology

Group presentations—guidance notes:

- Your group will deliver a formal presentation of the security problems you have found with the assignment text together with your proposals for their solutions.

- The presentation will be aimed at the senior management of the company who must not be assumed to be very computer literate.
- The presentation will last 15–20 minutes and there will be 5 minutes of questions from the management to the group.

Assessment, for which an individual mark will be given, will be made under the following headings:

Group (50%)

- Timing—The presentation should proceed at the appropriate rate with each individual contributing.
- Introduction—Someone should introduce the purpose and structure of the presentation and the members of the group.
- Continuity—There should be neat handovers from member to member.
- Conclusion—The audience should be told that the presentation has ended and that questions may be asked.
- Professionalism—How professional is the group and its performance?

Individual (50%)

- Performance—Are you articulate and confident?
  You should not appear just to be reading a script.
  Make eye contact with the audience.
- Content—Your presentation should be aimed properly at the target audience.
  It should be correct and understandable.
- Materials—You should use audio-visual aids (OHPs, handouts, etc.).
  They must be easy to see and understand.
  You should use these properly (e.g. cover OHPs).
- Input—Your input should be about the same amount as the rest of the group.
- Questions—Replies to these will be assessed individually.

Soft systems techniques

☐ Hard and soft systems techniques
☐ Soft methodologies
☐ Soft Systems Methodology (SSM)
☐ The seven stage model
☐ SSM in information systems development
☐ PG TOPIC: Soft versus hard

5.1 Hard and soft systems techniques

Shelfware is the word for IS deliverables that stay on the shelf rather than being used.[8]

The methodological procedures we have discussed so far are often referred to as *hard* approaches to information system development (ISD). These techniques are sometimes said to be the sort of procedures carried out by those who have little care or understanding of the softer aspects of life, such as those involving human beings and their interactions within the organization that will host the information system (IS) being developed. The hardness of such methodologies has been seen as a reason why so many computer systems have become *shelfware*.

The argument is that hard ISD ignores the effect of new systems on the staff of the company and that the elicited requirements of such systems do not take enough account of people and personal interactions. This is what Flynn (1998, p. 333) refers to as 'the wider organizational context'. As a reaction to this negative view, some thinkers have proposed that systems development should be more soft rather than hard.

---

[8]   There is hardware and software; but also **wetware**: the sludgy stuff inside people's heads is wet—the brain. So *wetware* refers to the humans within the IS.

5.2 Soft methodologies

The important difference between hard and soft ISD is a change in viewpoint: from the machine solution to the organizational solution of the problem. This means spending much more time and trouble looking at the wider organizational context, rather than concentrating only or mainly on the way the computer can solve problems. This chapter aims to be an introduction to the major concepts and techniques of the specific version of soft methods developed by Peter Checkland and his colleagues.

The major name in the attempt to move ISD from hard to soft development is that of Peter Checkland. Checkland (1981) and Checkland and Scholes (1990) are the standard texts on what he calls **soft systems methodology (SSM)**. The methods described concentrate on finding out what is really wanted from a new IS by extensive interviews of the people in the sponsoring company. Greater importance is also put upon the way the new system will affect the company and the staff in that company. A good book on the practical use of these techniques is Patching (1990).

There is also the belief that there is a difference between the so-called **reductionist** view of systems and the **holistic** view.

Table 5.1 shows the major differences between these views.

Soft methodologies tend much more to the holistic viewpoint; they also stress that we live in a social world that is not inherently orderly. This is because social action and events cannot be completely described in reductionist terms. In our private and working life we actively bring prior meaning to events and objects. In fact, social action is only possible when these meanings are shared and interpreted in a similar way. Soft issues need to be resolved by negotiation, persuasion, compromise, and agreement by consensus, rather than by dictate and the imposition of the management position.

Along these lines, Harry (1997) compares what he calls **hard problems** with **soft messes**. Hard problems are clearly defined and well-bounded; their information needs are known, as is the way the solution should look. It is also clear who ought to be involved in them. On the other hand, soft messes are undefined and fuzzy edged; there is doubt about what information is needed, what the solution would look like, and even who ought to be involved in them. Harry writes that most real problems fall somewhere between these two extremes.

Reductionism the belief that complex data and phenomena can be explained in terms of something simpler.

Holism the theory that the fundamental principle of the universe is the creation of wholes, ie complete and self-contained systems from the atom and the cell by evolution to the most complex forms of life and mind; the theory that a complex entity, system, etc, is more than merely the sum of its parts.

Chambers CD

Table 5.1 Holistic versus reductionist ideas.

Holistic	Reductionist
The whole is greater than the sum of its parts	The whole exactly equals the sum of its parts
Emergent properties	
Subjective	Objective
Soft system perspective	Hard system perspective
System and humanistic elements which are viewed as a system	Systematic and methodical

5.3   Soft Systems Methodology

SSM developed in the 1970s and grew out of the perceived failure of the more traditional development methods of systems engineering (SE) when faced with messy and complex real-world problem situations. The argument is that SE is concerned with creating computerized ISs to meet well-defined objectives. SE works well in those situations in which there is general agreement on objectives to be achieved, so that the problem can be thought of simply as the selection of efficacious (producing the desired effect) and efficient means to achieve them.

SSM provides a set of guidelines for examining an organization with a view to clarifying where improvements may be needed. It does not require strict adherence to procedures or rules, although there are certain constructive and strategic rules that assist with its application. Also it makes an explicit distinction between real-world and system-world activities. However, many of the actions taken by an analyst using SSM are conventional fact-finding activities.

The steps of SSM can be used in any order the analyst requires, and the method encourages the process of iteration as the analyst's knowledge increases. It advocates examining situations from different viewpoints, and establishes the basis for a debate with the client about possible changes. It is a participative approach, but it can still be of value even if this participation is limited.

SSM has been in use in industry since the 1970s, where it has tackled a wide range of problem situations from charity management to warehousing, as well as for industries from aerospace to telecommunications. SSM claims that it tends to be very well suited to ISD.

SSM is likely to have deliverables concerned with trying to define the nature of the problem itself, rather than offering complete solutions. It is useful in trying to obtain clarification and agreement of what needs to be done, and it is likely to be concerned with asking 'Where are we now?' and 'Where should we like to be?'

5.4   The seven stage model

Professor Peter Checkland developed this in the 1970s. The model asks 'What shall the situation be deemed to be?' It emphasizes that people have different understandings of many situations. The term for this outlook on the world is **Weltanschauung**, the person's world view (we shall meet this later). SSM looks at problems as intellectual constructs, and asks 'What constitutes a problem?' Its aim is to improve the problem situation, but accepts that it is subjective and that the analyst is part of the problem situation.

Figure 5.1 briefly lists the seven stages in the model. They will be discussed in some detail below.

5.4.1   Stage 1: problem situation: unstructured

When first encountering the problem, the analysts should resist the tendency to impose structure upon the situation. They should recognize that once they

> Stage 1:   Problem situation: unstructured.
> Stage 2:   Situation analysed: rich pictures and ideas about relevant systems.
> Stage 3:   Root definitions of relevant systems evaluated by CATWOE criteria.
> Stage 4:   Conceptual model of the system, built by assembling and structuring verbs (usually expressed in diagrammatic form).
> Stage 5:   Agenda for possible changes (derived by comparison of conceptual models with the rich picture expression of the problem).
> Stage 6:   Changes judged with actors in the situation to be (systematically) desirable and culturally feasible.
> Stage 7:   Implement the changes.

Figure 5.1   The seven stage model.

become involved, they become part of the very situation they are investigating. This means that their own objectives, biases, and reasons for becoming involved should be examined and kept in mind. Important also will be the various roles of the people involved: the **client**, the **problem solver**, and the **owner** of the problem.

5.4.2   Stage 2: problem analysis: rich pictures

Rich pictures are a way of summarizing and communicating significant aspects of a complex situation in an efficient, economical, and illuminating way. **Pictures** of what is the case are always useful as a way of communicating between participants; they are called **rich** because they can contain a large amount of information.

The development of rich pictures should look for processes and structures, and the way they interact. The relationship between these should give an idea of the climate within which the problem is situated. It is important—particularly for those who are inclined to look for computerized solutions—not to try to represent the situation in terms of systems. What is essential is to look for people playing the different roles of client, problem solver, and problem owner. The rich picture should include both hard factual data and soft subjective information. It can be annotated with footnotes, and the analysts should also be included in the picture: with their role and relationships. The generation of the rich picture will of course be a collaborative and iterative process.

As an example, Figure 5.2 is a rich picture of a hotel-based problem.

The rich picture is based around the front desk of the hotel. There are two participants, the head receptionist and the reservation clerk. The text close to the stick figures of these people, rather like speech bubbles in a cartoon, shows some of their concerns. Outside the front desk, rather like the environment of the system shown in Chapter 1, are the people whose jobs impinge on, or are affected by what happens at the front desk. Also shown are the guests of the hotel, who first experience the company through the front desk.

Example of a Rich Picture:

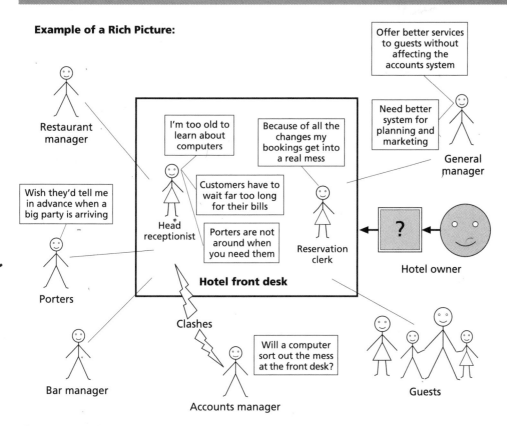

Figure 5.2 Rich picture for a hotel.

5.4.3 Stage 3: root definitions of relevant systems evaluated by CATWOE criteria

CATWOE is an acronym where the letters refer to the following: Customer, Actors, Transformation, Weltanschauung, Owners, and Environment. The terms are defined thus:

CATWOE

Customers: Those who are directly affected by any change (transformation) to information in the system. They may be considered to be beneficiaries, or even victims. In the rich picture above, the customers of the front desk are not only the guests, but also the porters, bar manager, and all those other staff affected by what happens there.

Actors: These are the people who carry out the changes. For the hotel front desk, the actors are at least the head receptionist and the reservation clerk.

Transformations: The changes to information in the system. This refers to what the system does with its inputs to turn them into outputs. For the front desk, transformations include what happens to the information taken from guests and sent on to the other customers, such as the restaurant manager and general manager.

Weltanschauung: The world view that makes these transformations meaningful, and puts them into context. This would include the whole culture of the particular hotel, and the hotel trade in general.

Owners: These are the people who have the power to stop any transformations. For the front desk, this would appear to be the hotel owner.

Environment: Refers to the constraints that impinge upon the system from outside, and are taken as given.

For a medical records department, the CATWOE analysis could provide the following results.

CATWOE for a medical records system

Customer: The clinical staff.

Actors: The medical records staff.

Transformation: Information such as treatments for patients are turned into patient history records that are made available to the clinical staff.

Weltanschauung: That which is generally expected of health care in the community.

Owner: Presumably the National Health Service (NHS) or the particular health trust.

Environment: One of the powerful external constraints for such a system is the fact that demand in a health system generally outstrips the funds available for its supply.

CATWOE for the hotel front desk

As another example, the hotel front desk case leads to a CATWOE analysis that fits the rich picture shown previously:

C: The front desk and other relevant employees of the hotel

A: The customers and employees

T: Guest needs and front desk needs identified and satisfied

W: Offer better services to guests without unnecessary delays

O: The hotel owner

E: Better relationships between different staff

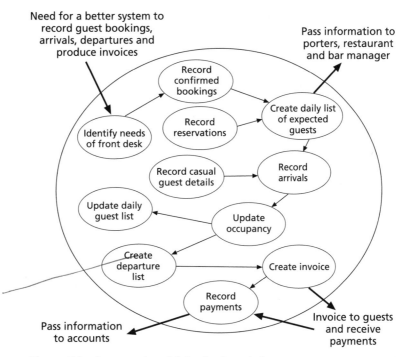

Need for a better system to
record guest bookings,
arrivals, departures and
produce invoices

Pass information to
porters, restaurant
and bar manager

Pass information
to accounts

Invoice to guests
and receive
payments

Figure 5.3 Conceptual model for the front desk.

5.4.4 Stage 4: conceptual models of the system

The soft systems **conceptual model** is a diagram of the activities in the system in which we are interested. You may feel that the example we give in Figure 5.3 has several similarities with the data flow diagrams we shall discuss in Chapter 8. You will see that the model is based on the use of verbs that refer to the activities needed to be carried out by the people involved in order to carry out the transformations in the system. Another name for such a model is a **human activity system**.

5.4.5 Stage 5: agenda for possible changes

The agenda for the changes that are needed in the system is worked out by comparing the rich pictures of the system with its conceptual models. The differences and similarities between the two models are investigated to give ideas about what is needed to improve the situation. These ideas form the basis of the debate that will be carried out among the participants to agree the next steps to take to make things better this is Stage 6.

5.4.6 Stage 6: changes judged with actors in the situation to be (systematically) desirable and culturally feasible

The debate should take place between, at least, the client, the problem owners, and the problem solvers. The aim of the debate is to find solutions that are both desirable to those concerned, and feasible given the cultural situation within which the *system resides*.

5.4.7　Stage 7: implement the changes

The final stage is to implement the agreed changes. The way this is done should, again, be agreed between all those concerned. It is important that putting the changes into effect is done in such a way as to bring about the benefits of the changes without leading to dissatisfaction of the participants involved.

5.5　SSM in information systems development

As presented so far in this chapter, it may be hard to see how a soft systems investigation leads to computerized ISs. Actually, SSM can be seen as being independent of the automation or otherwise of the final solution. There is no necessity for computers to be involved in that solution at all. Yet SSM is known mostly for its effects on IS developers rather than on people interested in other ways to solve system problems. For the purposes of this book, we are particularly interested in computerized ISs, so we must consider how SSM can be used in such an enterprise.

Several researchers have considered how SSM can be used to provide some input to ISD. Generally, they have thought that the importance of SSM is at the front end of the ISD process. Its major use is in eliciting the requirements of a new system. SSM concentrates on what the system participants want from a new system, so it can be seen as another way of finding out just what it is that they want. SSM stresses the importance of including all the participants in a system rather than only the owners or managers. We also stressed that it is important to talk to the right people, but SSM emphasizes that all the participants should be involved in deciding what the changes need to be and the way they ought to be implemented. Traditional management concepts might well suggest that it is the managers' and owners' jobs to make such decisions rather than also including the views of the ordinary staff, but SSM is all-inclusive. This aspect is very similar to the notions of Mumford's (1983) ETHICS approach, in which all staff are included in discussions of what is best to be done and the best way to bring it about. The aim of these sociological views is to try to ensure that the changes that take place in a company are actually the ones that the company and its staff require; and also that these changes are much more likely to be implemented with as little staff resistance as possible.

Even if it is agreed that the discovery of the requirements of the system can be addressed by soft system development, SSM says very little about how such changes can be brought about for computerized ISs. The main work on the knitting of SSM with ISD has been that of Multiview. Avison and Wood-Harper (1990) developed the Multiview idea that suggested that SSM could be used to define the requirements of a new system, as well as create staff acceptance of the changes that would be needed to improve the situation and solve its problems. They carry on to suggest that these requirements should be fed into the rather more traditional IS development procedures that this book has already described briefly, and that will be dealt with in much more detail later.

The Multiview concept therefore tries to attach the soft systems techniques to the hard system techniques generally used to develop a computer system. Briefly, SSM can be used to elicit the system requirements and hard techniques to develop the computer system.

5.6   PG TOPIC: soft versus hard

The use of the terminology in the argument of soft versus hard methodologies is interesting to investigate. The word *hard* does not seem to have the same meaning as *difficult* or even as in *hardware*, but rather it seems to have more to do with its meaning in *hard*-nosed, *hard*-boiled, and *hard*-hearted. Continuing the derogatory aspects of the word, traditional IS personnel may also been seen as *hard*, and so having less interest in the human aspects of IS than they ought.

Those of us who have been involved in the computing profession for a long time do not recognize this caricature of our work in the past, but there is no doubt that many computer systems have been developed that do not do what they need to—ending up as shelfware.[9] One of the problems with the design of these unused or unusable systems has been seen to be their lack of consideration of the human aspects of systems analysis and design. The advanced student may like to consider how fair is this description of the development of ISs from the 1960s.

It is doubtful that the apparent lack of interest of IS developers in all the views of anyone who might be seen to be involved in some new system is very much different from the way managers in business normally act. A manager might be expected to take account of the feelings of staff before introducing some innovation into the workplace. However, no manager would be likely to spend so much time on finding out about these views that the proper time for the new introduction is allowed to pass. There comes a time when the manager's job to manage means that he or she must take the responsibility of making some change because they believe that it will be in the best interests of their company.

In real-world practice, IS developers are constrained in the way they elicit requirements by the management of that development and the management of the departments for which that development takes place. In the corporate world, IS developers are often employees like any others, or brought in by corporate management in order to implement their view of what is required. Any change in this situation is the domain of the sociologist or the political activist, not those trying to earn an honest crust by doing what they have been paid for doing.

One of us (Chester, 2000) presented a very short conference paper on the intrusion of sociology into ISD and was severely criticized by some of those present, who appeared to think that the whole of ISD is (or should be) completely subsumed into the realm of sociology. The final remark to the author by one of the organizers of the conference was that 'We are all sociologists

---

[9]   *Shelfware* is software, or a whole system, that is never used and so remains on the shelf.

really'. Needless to say, this is not the view of the writers of the book you are now reading. However, we have no doubt that people are extremely important for ISD and their views must be taken into account when deciding what ISs are required. This is, after all, the whole thrust of Chapter 3.

Our position is that the systems that are to be developed are those that the responsible people in the company say that they want and need, rather than presenting them with systems that are acceptable on someone else's ethical, sociological or political opinion. For the politically committed thinker, such a position is completely unacceptable. To their mind, the politics or sociology comes before the stated needs of the managers of a company. If this is your view, then it is perfectly acceptable, but may require the payment of a high price. For all of us, there may come a point where the system that the management insists upon being instituted is so politically or ethically repugnant that we feel we cannot be involved in it. In that case, it is our moral responsibility to argue against it, and, if necessary, resign rather than be involved. But that is no different for the development of a new computerized IS than for any other business change.

Entity relationship diagrams I: logical data structures

6.1 Entities

We shall now embark upon the tricky concepts of entities and the relationships between them. Some definitions of what we are talking about would be a good idea, but, sadly, they are not readily available. An entity is at least an **object** or a **thing**, but not just any *object* or *thing*. The box is one attempt at defining an *entity*.

However, this definition leads to the questions of what is an object, and what makes it of interest? These questions are also not easy to answer. Perhaps the best way to find these entities or objects is to ask what information is being (or needs to be) held or handled by the IS. Or, better, to ask what the information is being held **about**. If the IS holds (or needs to hold) information about customers, then

> *An entity* is an object in the world that is of interest to the information system under consideration.

CUSTOMER will be an entity for that system. It may also hold information about orders and suppliers. If so, then ORDER and SUPPLIER will be entities.

To be fussy, the above are really examples of *entity classes*. Some would say that an entity is a **particular** customer or order or supplier. The whole set of customers, about whom we hold information, is the entity class CUSTOMER. We shall be less fussy and speak about *entities* rather than their classes, but you should be aware of this technical distinction.

6.2   Relationships between entities

Entities have relationships with other entities, because data about one entity has associations with data about another. For example, the information held about our customers will obviously relate to the data about our orders. Somehow, we need to be aware that a particular customer placed a particular order. Also, more generally, a customer is allowed (and encouraged) to place several orders; while (probably) a particular order is placed by one—and only one—customer.

In the parlance of entities and their relationships, a CUSTOMER may place one or more ORDERs; and an ORDER is placed by one and only one CUSTOMER. There may be systems where more than one customer places the same order, but let's not worry about those at the moment.

Note, even at this early stage, that there are **two** relationships between two entities. One goes one way: from the CUSTOMER to the ORDER; and the other goes the other way: from the ORDER to the CUSTOMER.

> **The logical data structure** of an IS is the way data in the system relates to other data.

> **The ERD** is a model, in a diagram or picture, of the logical data structure of the system.

> **ERA diagram** Short for entity-relationship-attribute diagram. A diagrammatic notation for describing and documenting data items and the relationships between items.
>
> *DOC*

6.3   Entity Relationship Diagram (ERD)

This is one way of showing the **logical data structure** of an IS.

It is **logical**, because it is the conceptual view of that structure that is or will be made **physical** by the files of the IS.

You may come across the terms 'entity relationship model', 'entity relationship attribute model', or 'entity relationship attribute diagram'.

We shall use ERD for the diagramming technique used here, and *logical data structure* (LDS) for the concepts that such a diagram symbolizes. The techniques employed here use the forms set out in structured systems analysis and design methodology (SSADM).

6.4   A preliminary example

Rather than give you the individual elements of ERDs and work up from those, here is a simple ERD of relevance to the writers of books and the users of libraries. We shall discuss what it means below.

This diagram can be explained thus:

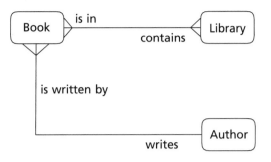

An AUTHOR writes one or more (i.e. several) books; and one BOOK is written by one and only one AUTHOR. Immediately you will spot a difficulty here: the book you are reading was written by more than one author. But, for the sake of this explanation, let us assume that only one author is involved. To continue with the diagram: a BOOK is in one or more LIBRARIES, and one LIBRARY contains several books.

The entities BOOK, LIBRARY, and AUTHOR are shown in boxes, and the fact that there is a relationship between two entities is shown by a line connecting those entities.

The words on, or lying beside, those lines are the relationship names of the relationship. Because there are two relationships between any two entities (one going one way and another the other way), there are two relationship names on either side of the relationship line.

The relationships **one or more** and **one and only one** are shown by the use of lines ending with a $\leqslant$, and lines without a $\leqslant$. This trident symbol is often referred to as a *crow's foot* on the relationship line.

You should now be able to see how the words in the explanation refer to the symbols in the ERD.

6.5   The ERD notation

The notation used for ERDs in this book will be from SSADM version 4+. We need to be able to draw entities, their relationships, and the degrees of those relationships:

An entity

The entity is shown as a box with rounded (so called soft boxes) corners and an entity name inside it—in this case A. Entities do not exist alone, they are related to other entities. The types of relationships are set out below:

1 : 1 relationship

If two entities A and B have a relationship, this is shown by a line between them. A line with no crow's feet at all represents that fact that for every A there is one and only one B, and for every B there is only one A. The degree of this relationship is called a *one-to-one* or *1 : 1 relationship*.

1 : *m* relationship (master–detail)

In this case, one A relates to one or more B, but one B relates to only one A. The relationship degree is called a *one-to-many* or *1 : m relationship*—this is also known as the *1 : m relationship*.

> **Master entity**
> Where two entities are connected by a 1 : *m* relationship, a single instance of one entity is related to several instances of another. The entity at the 'single' end is deemed the **master** entity.
>
> **Detail entity** ...
> The entity at the several end is deemed the **detail** entity.
>
> *Goodland and Slater,*
> *1995*

The crow's foot at the B end show that several Bs are involved for every one A. Having no crow's foot at the A end points to the fact that for every B there is only one A. Naturally, it is important that the crow's feet are put at the correct end.

The one-to-many relationship is also referred to as the **master-detail relationship**. In the above example, the A entity would be the *master entity*, while the B entity (at the crow's foot end) would be the *detail entity*.

This terminology seems to come from the idea of a master's relationship to slaves—generally, there is one master to several slaves.

m : n relationship

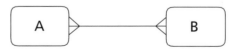

This is the highest degree of relationship and represents the *many-to-many* or *m : m relationship* between two entities. This diagram says that for every A there are several Bs, and for every B there are several As.

6.6 Entities and relationships

To reiterate:

- ENTITIES are objects in the real world that are relevant to the IS.
- Entities do not exist in isolation.
- Entities are associated by RELATIONSHIPS. e.g.
  A person is associated with one job title.
  A job title may have several people associated with it.
- An ERD is a pictorial way of modelling the relationships between entities.
- Both sides of a relationship have a RELATIONSHIP NAME.

6.7 Entities, attributes, values, and keyfields (identifiers)

Entities may be physical or conceptual. An entity may represent a customer or perhaps a building, but it could also be a project or a cost centre.

Attributes are the properties of an entity that are of interest to the IS. They are the individual items of data that are to be held about a particular entity.

For instance, the entity STUDENT may have the following attributes:

Student number, surname, initials, address, degree class, . . .

The entity EMPLOYEE may hold:

Employee number, name, address, job title code, . . .

Values of an attribute are the actual data for an attribute for one occurrence of an individual entity.

So, for a particular student in the STUDENT entity above, we may hold the following in the attributes described:

99506010, Chester, MF, Northampton, 2:1

In the Employee entity above, for an individual employee, we could have:

3342, Athwall, Wolverhampton, CEO

The **key field** (or **identifier** or **key**) of an entity is one of the attributes of an entity that is designated as the one whose values will uniquely identify the individual entity.

For the STUDENT entity, the key field is going to be the **student number**, and for the employee entity, the key field will be the **employee number**. Chapter 14, 16 and 17 go into more detail about keyfields.

Table 6.1   STUDENT (logical).

Student number	Surname	Initials	Address	Degree class
9856744	Athwall	AK	Wolverhampton	1
0085755	Chester	MF	Northampton	2 : 1
9756411	Evans	S	Coventry	2 : 2
9645388	Singh	A	Coventry	3
9900110	Smith	I	London	1
9876541	Smith	MC	Birmingham	2 : 2

Table 6.2   STUDENT (physical).

Student number	Surname	Initials	Address	Degree class
0085755	Chester	MF	Northampton	2 : 1
9645388	Singh	A	Coventry	3
9756411	Evans	S	Coventry	2 : 2
9856744	Athwall	AK	Wolverhampton	1
9876541	Smith	MC	Birmingham	2 : 2
9900110	Smith	I	London	1

6.8   Logical and physical data models

A **Logical data model** is a model of the IS's data that is not related to any physical method of storage. It is often the data as the user wishes to see it.

The ERDs we have discussed model entities and their relationships without being concerned about the physical way the data in the entities are stored. This makes the ERD a way to show the *logical data structure* of the IS.

For the STUDENT entity the logical data model might be shown as Table 6.1. The user may think of it as the list of students in surname order.

The **Physical data model** shows the data in a manner that suits the required IS processing. For example the system may hold the student information in student number order. This is shown in Table 6.2.

6.9   Developing the ERD

The task of producing the ERD (or the logical data structure) involves the following activities:

1. select the initial entities;
2. identify direct relationships;
3. construct the initial diagram;
4. determine the degrees of the relationships;
5. identify additional characteristics;
6. validate the ERD.

1. **Select the initial entities:** Remember that an entity is something significant to the system, about which information is or will be held. Also, an entity is uniquely identifiable. A baked bean is unlikely to be an entity because each bean is unlikely to be identified individually in the IS. However, a type of baked bean (such as plain, curried, or with sausages) can be.

   Also, there must be attributes that are related to the entity you choose. An entity with no attributes will be of very little use, so the selection of an entity will go along with the identification of its attributes or data.

2. **Identify direct relationships:** With several entities selected, look at them in pairs, and decide whether or not there is a relationship between each pair.

3. **Construct the initial diagram:** Draw each entity as a box and draw in the relationship lines between those entities that are related.

4. **Determine the degrees of the relationships:** Each relationship can be $1:1$, $1:n$, or $m:n$. We shall have to speak more about the last of these.

6.10 Many-to-many relationships

Although generally not strictly a part of the systems analysis of the current physical system, it is common at this stage to consider the special case of many-to-many relationships.

A $m:n$ relationship means that one item of one entity relates to several of the other entities and vice versa. Our example above was that a BOOK is in several LIBRARYs and a LIBRARY contains many BOOKS. Also, a stock PART will be found listed on several ORDERs, and an ORDER will contain many PARTs. Such relationships are very common in the world, and there is therefore nothing wrong with them. However, looking forward to the time when we would like to produce a computerized version of this IS, it is good practice to consider the logical and computing consequences of the physical fact that many-to-many relationships exist.

If we were to hold details of every BOOK in our IS, and also every LIBRARY, the $m:n$ relationship between them would suggest that, in order to relate a book to its library, we should have to hold a library name against each book; and also (for similar reasons) hold every book against each library's details.

Some current physical systems will work like this. If asked in which libraries a particular book is to be found the book's details must be looked up, and a list of libraries (perhaps a long list) read from them. If the question is posed as to what books you can find in a particular library, the list (a very long list) would be read off from that library's details.

Now consider what must be done in order to add one book to several libraries. The book details will be added to the BOOK entity, together with the list of libraries in which it is to be found. Also, the details of each library involved

---

[10] In Thomas Hardy's *The Return of the Native* (1878), a *reddleman* appears. Weirdly, he actually sells the colour red (redding or ruddle)—for marking sheep and such like. In his case, the colour red is his whole product.

must be looked up in the LIBRARY entity and the book's details added to each one. This laborious task would be much simplified by the use of a *cross-reference list*.

6.10.1 Cross-reference list

Many users of paper-based ISs have realized that this is a cumbersome procedure and have generated the very useful facility called a cross-reference list. This is a list of all the keys of one entity listed against the relevant keys of the other entity. In the case of the books, there could be a list of all the book numbers listed against the library names (or IDs) to which each book belongs. Adding a book to a library is then a comparatively simple matter of putting the few entries into the cross-reference list of the book against the libraries to which it is being added.

The cross-reference can be then used to answer simple questions about which book is in which library and what libraries contain which books. Cross-references are very common in current physical systems, having often been produced because of the frustration of the users of the IS with the natural $m:n$ relationship.

The full details of each book are held in the BOOK entity, and the complete information about a library is held in the LIBRARY entity. In a computerized IS for this system, the books would probably be sorted in book number (ISBN) order, and the libraries would need some type of ID or library number. A cross-reference list for this could then consist simply of a table with each book number followed by its relevant library number, as in Figure 6.1.

Note that a particular book number may appear against two or more different libraries, and a particular library number will have several different books listed against it. This is very common in any cross-reference, so that it can be used to look up a particular book to find out in what libraries it is to be found. A similar cross-reference, sorted in the order of the library number (as in Figure 6.2) could be used to discover what books are to be found in which libraries.

6.10.2 Link entities

So, sometimes cross-references really exist. However, if they do not, then for logical and computing reasons we need to use something very like them to cope with the $m:n$ relationship between two entities.

Book Number	Library Number
124 124 243 5	L124
135 757 35 568	L124
234 2345 4634	L451
234 2345 4634	L101
234 42324 4 1	L124
234 42324 4 1	L101

Figure 6.1 A Book–Library cross-reference in book number order.

Library Number	Book Number
L101	234 2345 4634
L101	234 42324 4 1
L124	124 124 243 5
L124	135 757 35 568
L124	234 42324 4 1
L451	234 2345 4634

Figure 6.2 A Book–Library cross-reference in library number order.

Even if it does not exist physically, we can handle the concept of the cross-reference in our ERD. In logical (non-physical) terms, we can invent an entity that joins the original two entities, turning one $m:n$ relationship into two $1:n$ relationships. The entity so invented is called a *link entity*.

To use the library book example:

Using a link entity, this becomes:

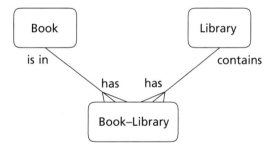

There seems to be a tradition of naming the link entity with the names of the linked entities separated by a hyphen. So, the link entity between BOOK and LIBRARY becomes BOOK–LIBRARY. (The alternative, LIBRARY–BOOK, could be confusing). If possible, a more meaningful name should be found. In this case, LOCATION would perhaps be better.

Inspect what has happened here. A new entity has been invented (assuming a cross-reference list does not already exist), and joined to the original entities by relationship lines. Now, instead of the crow's feet being on both ends of the line attached to each original entity, both the crow's feet have migrated to be on the link entity. The result is that we have two $1:n$ relationships instead of one $m:m$ relationship. Our cross-reference list has been brought into being, at least conceptually.

Again, in a computerized IS the individual elements of the BOOK entity will have unique **book numbers (ISBNs)**, and those in the library entity will have unique **library IDs** or numbers. The BOOK–LIBRARY link entity has a key of the **ISBN** combined with the **library ID**.

6.11 Example 6.1

This is a fairly straightforward example of an ERD. But watch out that you model **exactly** what is said, and not what you think ought to be the case.

A country bus company owns a number of buses. A bus serves only one route, although on each route there can be many buses.

Each route consists of many stages, where a stage passes through a number of towns.

One or more drivers are allocated to each stage of a route, but each driver is allocated to only one stage.

A stage belongs to only one route. Many stages can pass through each town.

From the above (you might think odd), description, draw an ERD to show the corresponding entities and relationships.

We shall give you all the entities first. See if you can draw in the relationship lines and the degrees of relationship. If you find a many-to-many, keep it as it is for the moment.

Do not forget the relationship names, two for each relationship line.

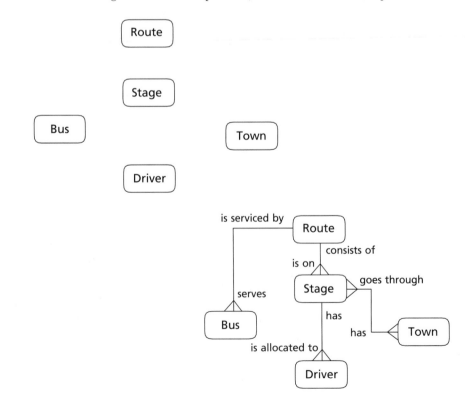

Then decompose the many-to-many:

STAGE–TOWN is the invented *link entity.*

Note how the crow's feet move from the original entities to be attached to the link entity.

6.12 Tutorial 6.1

Draw the relationships, degrees, and add relationship names for the following. There is room for discussion about the definitions of the terms used here.

1   a type of baked bean AND manufacturer;
2   a type of baked bean AND can label;
3   a student AND a taught module;
4   a student AND a university course;
5   a student AND type of beer;
6   a head of school AND school;
7   a head of state AND country.

6.13 PG TOPIC: What is an entity?

The academic literature and books on the subject tend not to deal with the question of the existence of entities. It is in fact very easy to come across situations where you cannot tell what a particular entity represents, or whether an individual member belongs in one entity or to another.

Let us take the example of the entity CUSTOMER. Naturally, many sales order ISs hold data about their customers. But is someone a customer before they have bought anything from you? Should there perhaps be another entity holding information about people who have not yet bought something from you—say POTENTIAL CUSTOMERs. Once they have bought something they could be moved to be a member of the CUSTOMER entity. However, who would you need to include as a POTENTIAL CUSTOMER?—the entire population? Or perhaps that part of the population who might be interested in your product? Or just those with whom you have had some contact?

How about the case of student APPLICANTs and enrolled STUDENTs? Are they one entity or two? In the student IS in our university—as far as we understand it—an applicant's application number is the same as his or her student number once they are accepted. Luckily, most of the time, the applicant is the same physical person as the finally accepted student. They seem physically to be the same entity, but logically they are different.

Entity relationship diagrams II: more complex relationships

7.1 Developing ERDs (continued)

As we noted before, producing an ERD requires going through the following steps:

> 1. select the initial entities;
> 2. identify direct relationships;
> 3. construct the initial diagram;
> 4. determine the degrees of the relationships;
> 5. identify additional characteristics;
> 6. validate the ERD.

We have dealt with steps 1 through 4, now to look at the last two.

5 **Identify additional characteristics:** There are several other characteristics to take into account before completing an entity relationship diagram (ERD) properly. We shall look at three: optional relationships, recursive relationships, and exclusive relationships.

7.2   Optional relationships

Relationships can be **mandatory** when they MUST exist (with which we have dealt so far), or **optional** when they MAY exist.

For example:

- A house **MAY** be let by **ONE** landlord (a house may be privately owned instead)

  BUT

- A landlord **MUST** let **ONE OR MORE** houses (in order to be called a landlord)

The MUST relationship is **mandatory**, and the MAY relationship is **optional**. The ERD for this would be:

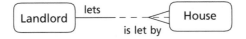

The dotted line at the HOUSE end of the 1 : m relationship denotes the fact the relationship of HOUSE to LANDLORD is optional; so one HOUSE MAY be let by one LANDLORD. The solid line at the LANDLORD end says that the LANDLORD : HOUSE relationship is mandatory—a LANDLORD MUST let one or more HOUSEs.

Again, it is important to put the dotted part of the line at the correct end; otherwise something different to what is intended will be depicted.

7.3   More on optional relationships

As we said, a relationship may or may not always exist. Looking at one in detail for the master–detail relationship of a borrower and a loan.

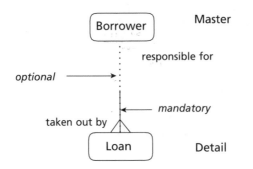

In English, this diagram represents the fact that:

- *each* BORROWER *may be* responsible for *one or more* LOANs; and
- *each* LOAN **must be** taken out by *one and only one* BORROWER.

As in life, there are two sides to any relationship, and each may be mandatory or optional. Two sides to a relationship and two possibilities for each make four possible combinations.

- mandatory at both ends;
- optional at Master and mandatory at Detail;
- mandatory at Master and optional at Detail;
- optional at both ends.

Optional relationships are also possible for one-to-one and many-to-many relationships.

7.4 Exclusive relationships

An *exclusive* relationship exists when the existence of one relationship prevents the existence of another.

If we assume, in a medical IS, that a patient can be either in a ward or an outpatient, **but not both**, the following ERD would result.

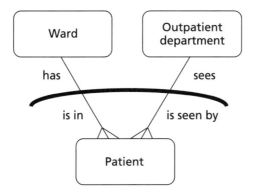

The strong line going across the relationship lines denotes that only one relationship exists, but not both.

So, a PATIENT is on one WARD

OR

A PATIENT is seen by one OUTPATIENT DEPARTMENT
BUT NOT BOTH

7.5   Recursive relationships

These are used where entity occurrences have direct relationships with other entity occurrences of the same entity type. A standard example of a recursive relationship exists where some EMPLOYEEs manage other EMPLOYEEs.

A manager is an employee; so employees manage other employees. This is a *recursive* relationship.

Taking into account the dotted part of the relationship, the diagram can be read as:

- an employee MUST BE managed by one (other) employee; and
- a particular employee MAY manage several other employees.

The circular relationship line depicts the recursive relationship. It is recursive because the relationship refers back to the same entity. This is often called a *pig's ear*.

6   **Validate the ERD:** As stated previously, every deliverable should be checked before the development of an IS is allowed to continue. In the case of an ERD, at least three things can be checked to validate it:

- No redundant relationship should be shown.
- The structure should be able to describe and support the processing requirements of the current system.
- The model must be acceptable to the user.

7.6   One-to-one relationships

1 : 1 relationships need some further words. Whenever a one-to-one relationship is discovered, they should be examined very carefully. For instance, in a company personnel IS the following may be discovered:

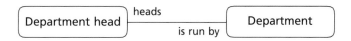

It would not be unusual for the IS to store details against the DEPARTMENT; and it may be that the system stores details about every DEPARTMENT HEAD. However, it is more likely that the DEPARTMENT HEAD (or at least an employee code for the department head) is stored as an attribute of DEPARTMENT. If this is the case then these two apparently separate entities are actually one entity.

The danger with 1:1 relationships is that we may be looking at only one entity, not two.

7.7 Review of entity relationship diagrams

Before an entity relationship model, or ERD, can be drawn, a list of entity types has to be made.

The entity types listed can then be put into a model and the relationships drawn.

The entity model can be developed and then modified. It might be drawn several times until it is correct.

7.8 Example 7.1

In a company database, the following details have been discovered. Draw the ERD. Take careful note of the plural nouns—they suggest that several of these attributes occur for the relevant entity.

Entity	Attributes
Employee	Employee number, name, manager, department, grade, room, current projects
Department	Department code, name, building
Building	Building code, name, location, size, departments, room numbers
Room	Room number, building, floor area, height, employees
Project	Project code, name, employees, client
Client	Client number, name, projects

A skeleton diagram for this would be:

And the completed, initial diagram would be:

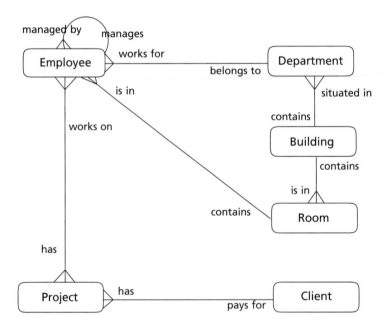

However, we also need to resolve the $m : m$ relationship between EMPLOYEE and PROJECT.

This should be broken down into two $1:n$ relationships.

So the above becomes:

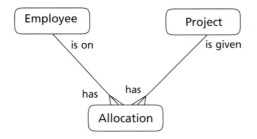

Here, ALLOCATION is a *link entity*; another suitable name for which could be EMPLOYEE–PROJECT. It is a cross-reference showing which employee is on which project.

7.9 Data relationships

At this stage it would be useful to be aware of the way the relationships between data, as represented by ERDs, are actually represented in a paper-based or computerized IS. This will also be of use as an introduction to the rather difficult techniques of relational data analysis (or normalization) that we cover in Chapter 14, especially to help show why we carry out such procedures.

To do this it will be necessary to discuss the way data must be held in order to represent the three degrees of relationship: one-to-one, one-to-many, and many-to-many.

7.9.1 One-to-one relationships

An IS containing a $1:1$ relationship, such as that which pertains between a department head and the department seen before, needs to hold a list of department heads and a list of departments. It also needs to be able to link each department head with his or her department. Without such a link, there would be no way to tell who heads which department. In a paper-based IS it would not be uncommon to do it as shown below.

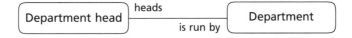

In one file there is a list of department heads, each one of which has a unique code or key field (the head code for each department head), along with the details of that head. Also, in a file of departments, there will be, for each one, a unique department code along with the details of that department. The key fields will be denoted here by a ∗.

At the moment we have not described how these two files are linked together. The links are provided by the fact that, as shown below, each department file entry contains a place to hold the department head number, and each department head entry has a department number. So, to find out the details of a head's department (knowing only the head code) one could look up the department head, find the department number, and then look up the details of that department. Conversely, a similar procedure could dig out the details of the head of a particular department by knowing only the code for that department. In the department file, the department head code is what we might call a *link field* to the department head file, and the department code in the head file would be a link field to the department file. We shall later learn to call such link fields, *foreign keys*. In the example, these are shown in italics.

One-to-one relationships

- In a database of companies and managing directors:

Company

* Company number				
Name				
Managing director code				

Managing director

* Managing director code				
Name				
Company number				

- *Managing director code* in **COMPANY** file points to the key field in the **MD** file.
- Could also have *Company Number* in **Managing director** file to point to the key field in **COMPANY** file.
- In a computer system, probably better to pick one solution only. Or combine them.
- In a paper-based system, you often find both.

One-to-one relationships example

Company

* Company number	C1	C2	C3	C4
Name	WH Smith	Railtrack	M&S	GEC
Managing director code	MD3	MD1	MD4	MD2

Managing director

* Managing director code	MD1	MD2	MD3	MD4
Name	Smith	Jones	Hacker	Jacks
Company number	C2	C4	C1	C3

In a computerized system it is usually unnecessary to hold link fields in both files. Given the ability of computers to sort files quickly, or at least their keys, it is generally sufficient to use only one link field.

7.9.2   One-to-many relationships

In a $1:n$ relationship between two entities, such as that between a department and its staff, two files must again be held. One file is sorted on pay number (with the details of that member of staff alongside) and another file is needed to hold details of departments sorted on department number. The way of linking these files is not the same as in $1:1$ relationships. It is not possible to hold one field for one pay number in each department's details. This is because there will be several staff in one department. In a paper-based system, a list of staff may well be held in this way, but for our purposes that is not acceptable. In fact, the only place we can link the two files is by holding the number of the relevant department against each employee. (A paper-based system might do that as well). This would give files like those shown below.

One-to-many relationships

MUST have the pointer field on the detail entity.

Department
* Department number
Name

Staff
* Pay number
Name
Department number

Department number on **Staff** file points to the key field on **Department** file.

One-to-many relationships example

Department

* Department number	D1	D2	D3	D4
Name		Accounts	Warehouse	Personnel	IT

Staff

* Pay number	P1	P2	P3	P4	P5	P6
Name	Smith	Jones	Singh	Singh	Hacker	Jacks
Department number	D2	D2	D4	D3	D1	D4

The link field (or *foreign key*) is the department number in the staff file. This would allow a computer system to find a department number if only the pay number is known. In the master–detail terminology, the link field is held on the detail entity, not the master.

7.9.3   Many-to-many relationships

The highest degree of data relationship, the $m:m$, needs to be treated carefully. As we have said, the traditional way of coping with these is to have a cross-reference list or a link entity. This is a file that holds mainly a list of key fields of one entity against the related key fields of the other entity. There can also be other fields held in this link entity that are related to each particular entry. In the case of the relationship between a warehouse and the parts it houses, the LOCATION entity (or file) is shown below.

In this, a so-called *compound key* is set-up, made up of the two keys of the original entities: the part number and the warehouse number. This is all that is absolutely essential to link the warehouse file with the part file so that, knowing only the warehouse number, we can find the parts that are stored there. The procedure also works the other way. Knowing only the part number, we may find all the warehouses that store it. However, there may be some data that is not just pertinent to a warehouse or to a part, but actually to a particular location. For instance, it is sensible to hold the information of how many of a particular part is held in a certain warehouse. The only place to hold such data would be in the LOCATION file in the **warehouse quantity** shown on p. 83. This data is not relevant only to the part, nor only to the warehouse, but to both. The terminology is that such data is *dependant* upon **both** entities.

Link entities

Warehouse		Part	
*Warehouse number		*Part number	
Warehouse name		Description	
List of Part Numbers		*List of Warehouse numbers*	

Need a link entity—LOCATION (or WAREHOUSE-PART)

Warehouse	Location	Part
*Warehouse number	*Warehouse number	*Part number
Warehouse name	*Part number	Description
	Warehouse quantity	

Many-to-many relationships example

Warehouse

* Warehouse number	WH1	WH2	WH3	WH4
Warehouse name	London	B'ham	W'ton	Bristol

Location

* Warehouse number	WH1	WH1	WH2	WH2	WH2	WH3	WH4
* Part number	P1	P4	P2	P3	P4	P3	P1
Warehouse quantity	200	400	10	500	300	210	330

Part

* Part number	P1	P2	P3	P4
Description	Nail	Screw	Nut	Bolt

7.10 Tutorial 7.1

7.10.1 Draw the entity relationship diagram

A company keeps Sales–Order information. Customers place a number of orders and an order can only arise from a single customer. An Order consists of an Order-header and a number of Order-lines. The Order-header contains Customer name, delivery address etc., and the Order-line states the part ordered and the quantity. Parts are stored in various warehouses and some parts are stored in several warehouses.

Add suitable relationship names.

Comment on the relationship that exists between Order-header and Order-line.

Data flow diagrams I: basic ideas

- Diagrams of data flows
- The initial example
- The components of data flow diagrams
- Data flows
- Data stores
- External entities
- Processes
- Tutorial 8.1

8.1 Diagrams of data flows

Data flow diagrams (*DFDs*) are a method of modelling, by the use of drawings, the processing of an information system (IS). They include the processes themselves (the activities or functions of the system), the information or data flowing into, out of, and within the system, and the way information is stored in the system.

DFD modelling is a tool common to several methodologies, but the version we are using is again taken from structured systems analysis and design (SSADM), like the way of representing entity relationships we have already used.

8.2 The initial example

Once again, we begin with a diagram to discuss, before embarking on the detailed description of the method. The example is of part of a video shop manual IS.

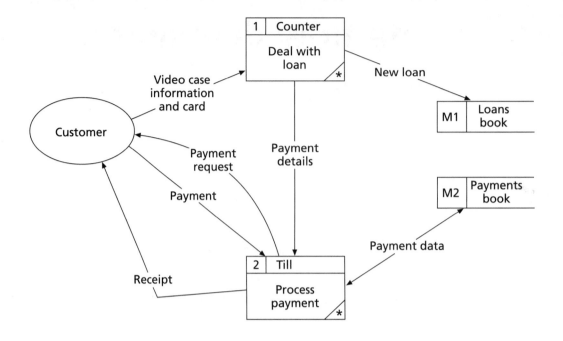

8.2.1 Video shop example Level 1 DFD

Without worrying too much about the details of the symbols used, it should be easy enough to understand that the video store customer or member takes a video case, containing the details of the video that is wanted, to the counter. Along with this comes the member's membership card. The counter deals with the loan by processing this information, and the details of the loan are entered in the loans book. In this shop, the details of the payment required are passed to the person at the till, and, in exchange for some payment from the customer, a receipt is issued to them. The customer is looked up in the payments book, and the details of the new payment are entered in there.

Some important things are worth pointing out even at this stage. All the arrows in the DFD refer to **information** or **data** moving (flowing) from one place to another. These are *data flows*. There is no flow for the physical videocassette or the video case. What flows from the customer to the counter, and informs the counter staff what video is required, is the **data** on the case. The only data that flows back to the member is the receipt for payment that may also contain details of the video rented and the date due back.

We shall give a way of showing the movement of physical objects, but the method is mainly one of data flows, not physical flows.

The payment data flow is interesting. The system needs to take account of the information that comes with a payment. In some systems only payment by cheque, credit card, or bank transfer is allowed. In a video shop, no doubt cash would be acceptable. Some of these movements appear to be flows of physical objects, but the IS treats them as data flows.

Components of data flow diagrams

Component	Symbol	Example
Data flow		Order
Process		2 Sales / Accept order
External entity		Customer
Data store		M2 Orders

Optional

Physical flow of		Parts

8.3 The components of data flow diagrams

There are only four symbols or components used in DFDs, plus one if you count the flow of physical objects, so there are not many things to remember. We show them here before discussing each component separately.

8.4 Data flows

Data-flow The relationship between a source of data and the repository or user of that data.

DOC

Data flows (or **data-flows**) show the information that is moving from one place in the system to another.

They represent data in motion. A data flow shows a pipeline or a channel through which data can flow between the other DFD components.

The flow is shown by a line joining the source and its user (the destination or *sink* of the flow). When data moves in the system, it goes in some direction, so the data flow has an arrow showing this direction. Sometimes data travels from the source to the user and back again (often after being modified). In this case a *bidirectional* data flow line is used with an arrow on both ends. If only one direction of the flow is needed, it is *unidirectional* and only one arrow is used.

In the video store IS, the data flow between the **till** and the **payments book** is bidirectional, because data flows from the till to the book and the other way as well. The process both reads the information in the **payments book** and also adds or changes data in it. This type of flow (both input and output) is usually used to show that, in order to change an entry in a place where data is held or stored, it is necessary to read it to find that entry first. The idea is that, in order to change the payment details, the employee must first look in the book to find the entry for the customer (perhaps ensuring that the customer is not in debt) and then add the latest payment to those details.

The name of the data flow appears on or next to the data flow line. A data flow name should be meaningful and unique.

Data flow names should be:

Meaningful: So that all those who need to know about it understand the name, including non-computer personnel.

Unique: So that one data flow will not be confused with others.

The new terms *data store* and *process* are yet to be described, but even now it is worth pointing out that a data flow always begins or ends with a process, or both. Data flows should not join data stores to data stores, external entities to external entities, nor any other combination of symbols that do not include at least one process.

Allowed dataflow connections

Between process and data store
Between process and external entity
Between process and process

Examples of data flows that might be found in the commercial world include orders, receipts, bank statements, payments, and copies of orders.

8.5 Data stores

While data flows are for data in motion, a *data store* (or *datastore*) represents data at rest. It shows a repository for information within the system, where a piece of data stays until it is needed to be moved elsewhere.

8.5.1 Parts of the data store symbol

The data store symbol has two parts to it. The identifier (or ID) space—the small box at the left (M1 and M2 in the initial example)—is for a unique reference for the store. This is so that a particular store cannot be confused with another. To help with this, the name space in the rest of the symbol identifies the store with a meaningful and unique name for the data.

We shall deal with this in detail later, but it is worth mentioning here that a data store generally relates to one of the entities in the entity relationship diagram (ERD) for the system.

Pitfall

Duplicated data stores have a double line on ALL occurrences (not just on the extra instances).

The double line appears at the far end (the left) of the symbol in the ID place, and **not** between the ID and the name.

8.5.2 Duplicating a data store

For clarity and neatness in the diagram, it is sometimes preferable to draw the same data store more than once. A double line is put at the left of the store's ID to show that the store is duplicated, thus:

If a data store is duplicated, then a double line is used as shown on **all** instances of the store.

8.5.3 Types of data stores

There are several types of data stores that can be used in a DFD. Data is stored in manual form or as a database (in computerized form). Information can also be held as transient data or permanently. The difference between transient and permanent data often causes problems. Another word for *transient* is *temporary*, and a temporary data store is a store that is capable, at times, of being completely emptied. Once emptied, the store then seems to disappear because it contains no data.

Manual data stores are generally paper-based, while databases are computerized. There can, therefore, be permanent manual data stores, permanent database data stores, transient manual data stores, and transient database data stores.

A permanent manual data store could be a filing cabinet, an address book, or a card index. It is a store that is paper-based, accessed by hand (hence, manual), and remains throughout the life of the IS (hence, permanent). The identifier (ID) for a permanent manual store is an M followed by a number. For example M1, M7, or M22.

A permanent database data store could be a computer file for employee records, a student file, or a stock file. It is a computerized store (hence, database), and remains permanently. The ID for this type of data store is a D followed by a number. Such as D1, D7, or D12.

Transient stores may be harder to find and spot—they are also harder to describe. A transient manual data store will be paper-based, accessed by hand, and will be emptied at various times throughout the lifetime of the IS (hence, transient or temporary). Two examples are common. There is the spike: a filament of metal, firmly attached at the bottom end to a block of wood or metal. Typically, a piece of paper is dealt with or generated by someone, who spears the paper onto the spike to be collected later. These were once often seen on the benches of workplaces, the counters of parts warehouses, and in newspaper offices.[11]

---

[11] In newspaper parlance, a news story that is written but never published is said to have been '*spiked*', because it has been put on a spike and never taken off to go to the presses, or has simply been dumped.

The spike transient data store could be shown in a DFD as follows:

TMI	Spike

Another example (perhaps more common) is that of the in-tray and out-tray system in use in many offices. Papers are put into an out tray so that at times the tray is emptied and the papers distributed to other people's in trays.

It is important to note that a transient store is completely emptied of its data at various times. The spike is completely emptied and so, at least theoretically, is the out tray.

An example of a transient database data store is a file that is produced as a holding store for data that is part way to being sorted. This is often called an intermediate sort file or a work file. At the end of the sort the file is completely emptied or the whole file is deleted.

The identifier for a transient manual data store is TM (or T/M) followed by a number: TMn. The transient database store identifier should be TD (or T/D) and a number: TDn. Tn(M) and Tn(D) might also be seen. Examples are: TM1, TM10, TD1, and TD8; or T1(M), T10(M), T1(D), and T8(D).

Types of data stores		
Type	*ID*	*Example*
Permanent Manual	M1	Filing cabinet
Permanent Database	D1	Employee file
Transient Manual	TM1	Spike
Transient Database	TD1	Intermediate sort file

The LOAN data store in the initial example is a permanent manual data store; so it has been given the ID of M1.

8.6 External entities

Data comes into the system and goes out of the system. In a DFD, the components from which data comes (the data sources) and to which data goes (the sinks) are called *external entities*. They are shown as ellipses (squashed circles) like the CUSTOMER external entity in the initial example.

External entities are very similar to the entities in ERDs. They are objects that are of interest to the IS. From their point of view, they either supply data (as sources) or receive data (as sinks) from the system.

So external entities are shown as ellipses containing a name which is meaningful and unique.

8.6.1 Duplicating an external entity

For clarity in the diagram, external entities may be duplicated like data stores. Again an extra line or bar is used on **every occurrence** of a duplicated entity, thus:

8.7 Processes

Processes in a DFD are the places where activities happen within the IS. A process always has at least one data flow entering it and at least one data flow going out of it. This is where the data is altered or transformed.

8.7.1 Process identifier

The process box is split into three parts. The top left-hand part contains the identifier of the process box. This ID is a number, sometimes followed by other numbers and separated by full stops. It is not only unique, but it needs to follow a system of identification that we shall soon explain.

8.7.2 Location

The longer part, next to the ID is where the location of the process is put. The location is the place where the activity is carried out. It may be the name of a department, or a desk, or even the name of an employee.

8.7.3 Process name

The largest part of the process box is for the name of the process. As with other names, the process name should be meaningful and unique. Furthermore, traditionally, the process name is made up of an imperative verb followed by the rest of the activity. In the initial example, the process with ID 1 has a location of **Counter**, and a process name of **Deal with loan**. The process with ID 2 has a

Hint

Verbs are doing words. Imperative verbs are orders (Do this! Do that!)

Verb: *a part of speech consisting of a word or group of words that signify an action, experience, occurrence or state.*

Imperative *[verb]: the form of a verb expressing command, advice or request.*

Chambers CD

Useful imperative verbs for DFDs are: Process, Deal with, Carry out, Handle, and Perform.

location of **Till** and a process name of **Process payment**. In these, *Deal with* and *Process* are imperative verbs.

8.8　Tutorial 8.1

8.8.1　Mario's pizza parlour

This is a fast food restaurant that specializes in pizzas. On entering the restaurant, a customer sits down at a table and consults the menu. After deciding on her preferred dish, she gives her order to a waitress, who writes it down on a two-part ticket, adding the price as indicated on the menu. One copy of the ticket is given to the customer, and the other is passed to the chef as an instruction to prepare the order. At the end of the meal, the customer checks her ticket with the menu prices, and hands it to the cashier near the exit together with her payment. The cashier puts the payment in the till, gives the customer the till receipt, and puts the copy of the order on a spike.

1　What are the external entities to the system? (Hint: consider the chef as external to the system; what is the other?)
2　What are the main processes (actions) in the system? (We only want two)
3　What are the main data stores (files) in the system? (Hint: a spike is a sort of file; where else is information stored?)
4　What are the data (information) flows between the above? (e.g. the customer gives a verbal order to the waitress, and eventually receives a receipt from the cashier)

Try to draw the DFD.

Data flow diagrams II: levels of data flow diagrams

- Levels in data flow diagrams
- Level 1 DFD
- Context level DFD
- Functional decomposition
- Balancing DFDs
- Lower levels of DFD
- Developing the DFD hierarchy
- Stopping the decomposition
- Developing data flow diagrams
- Checking the DFD
- Tutorial 9.1
- PG TOPIC: Questions about DFDs

9.1 Levels in data flow diagrams

In any reasonably complicated information system (IS), it will be impossible to draw all the necessary details as one diagram on one sheet of paper. The methodologies that utilize data flow diagrams (DFDs) recognize this problem, and solve it by using different *levels* of DFD. This is done by producing a hierarchy[12] of diagrams, from Context, to Level 1, Level 2, and so on.

Dictionaries do not seem to have caught up with the particular meaning of *hierarchy* that we are using here. When you see the structure of the DFD

---

12   *Hierarchy* has a strange origin. It comes from the Greek word *hieros*, meaning 'sacred', and seems to relate to the grading structure in church government.

Hierarchy a body or organization classified in successive subordinate grades.

Chambers CD

hierarchy, you should be able to see how it looks very much like a management hierarchy diagram in a company.

9.2 Level 1 DFD

The initial example at the beginning of the last chapter is a Level 1 DFD. This gives the general overview of the IS. In it are to be found the main IS processes, data stores, and the data flows between them. It also shows the data flows between main processes and the external entities of the system. The Level 1 may also be called the *top level DFD*, but in fact there is a higher level, which we discuss below.

We repeat the Level 1 DFD for the video store here.

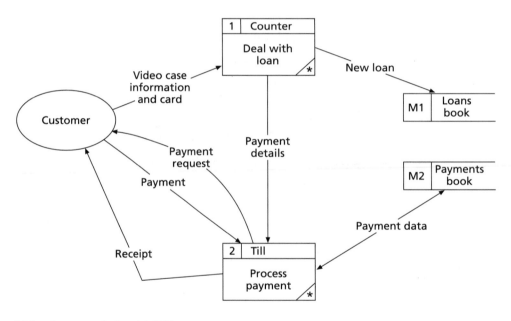

Video shop example Level 1 DFD.

9.3 Context level DFD

A higher level of DFD and the topmost level of the hierarchy for a set of DFDs is called the *Context Level* DFD. It may also be known as Level 0, but we shall not use this term. It shows the way the system interacts with its environment by showing the IS as a black box, its external entities (sources and sinks), and the data flows that run between the system and its sources and sinks.

The Context level DFD for the video store is easily derived from the Level 1 DFD above. We give it here.

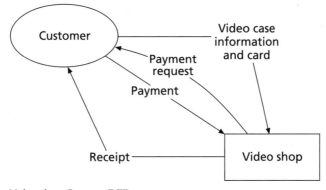

Video shop Context DFD.

You should be able to see how much this looks like the general system diagram given in Figure 1.1. The IS (**video shop**) is the black box and the data flows are the inputs and outputs. The addition of the external entities in the DFD makes it clear where the inputs come from and where the outputs go to. Note that there are no data stores in the Context level DFD.

9.4 Functional decomposition

The Level 1 DFD is like the subsystem diagram in Figure 1.2. It is important to note that the external entities in the Level 1 are exactly the same as those in the Context. Also, the data flows going to and from those external entities are exactly the same, with exactly the same names.

In the functional hierarchy, the Level 1 DFD is said to be at a *lower level* than the Context. The Level 1 depicts what is inside the black box of the IS in the Context, and expands upon it. The terminology refers to the *expansion* of a DFD at a lower level, or its *explosion* into more detail. The process of doing this is called *functional decomposition*. This is because the function of the system is shown in a more decomposed (or more detailed) state than it was at the higher level.

The word *decomposition* in *functional decomposition*, is used in the same sense as in program decomposition.

This harks back to our definition of *systems analysis*; *functional decomposition* is the *analysis*, at greater and greater detail, of the IS's functions or activities. It is carried out to break a complicated system into simpler subcomponents.

> *Decomposition* In programming, the analysis of a problem into simpler subproblems.
>
> DOC

9.5 Balancing DFDs

Two levels of DFD are said to be *balanced* when they have exactly the same output and input data flows (with exactly the same flow names). These flows should also go to and come from exactly the same sources and sinks for both levels of the DFD.

In the case of the Context level and Level 1 DFDs the only sources and sinks for each diagram are the external entities. We shall see that, for lower levels of DFD, other components must also be taken into account.

9.6   Lower levels of DFD

Successive levels of DFD show the system in more and more detail. They are called Level 2, Level 3, and so on.

A Level 2 DFD is a functional decomposition of one of the processes in the Level 1 diagram. It will have the same data flows (and flow names) coming in and going out as the Level 1 process. Here is the Level 2 DFD for the Level 1 process with ID **1**, carried out at the **counter**, and with the process name **Deal with loan**.

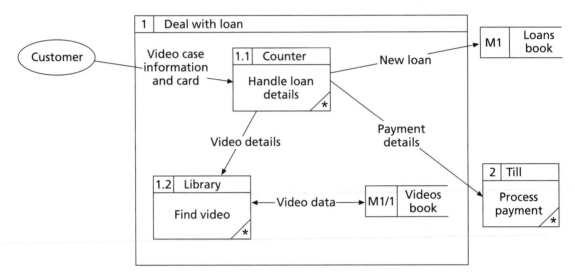

Video shop Level 2 DFD for process 1.

Look at the boundaries of this diagram carefully. You should notice straight away that it interacts with exactly the same external entity (**Customer**) as the Level 1 process. This part of the IS interacts with **Customer** through the data flow **video case information and card**. This is precisely the same as you will find for the Level 1 process 1.

However, the Level 1 process 1 also has a data flow (**new loan**) that puts data into the manual data store called **Loans book**. So in order to balance the Level 2 DFD with its Level 1 process, we have included the output and sink for the loan data as well. The sink is the **Loans book** data store, and the output is the data flow **new loan**. To be balanced completely, process 2 and the data flow **payment details** are also needed.

The point of decomposing process 1 is to uncover the more detailed procedures that allow it to operate. This is similar to finding the subsystems of the general system described in Chapter 1. So we need to include the Level 2 process that deals with the customer. In this case, the **counter** location has a

Pitfall

At Level 2, students often put data stores interacting with the Level 1 process INSIDE the boundary for the Level 2 DFD— this is wrong. Stores that are outside the Level 1 process that is being decomposed, stay *outside* the boundary for the process at Level 2.

Note that, in our example, the only store INSIDE the Level 2 boundary is a new one (**videos book**) that does not appear on the Level 1. The data store **loans book**, which was outside the Level 1 process 1, is also OUTSIDE the boundary of the Level 2.

process called **Handle loan details** to take the video case and membership card from the customer, record the loan in the **loans book**, and pass **payment details** to the **till**.

Also at Level 2, we find that another procedure takes place to find the video cassette and enter the fact that it has been taken out into another book. From the **Handle loan details** process, the particulars of what video has been ordered (**video details**) are passed to the **library**, whose task is to find the video. Once this has been done, the library staff look up the details of the video and enter the loan of that video in the manual data store called **videos book**.

No doubt the physical video is somehow given to the customer. But that contains no data of interest to the IS, so it does not appear here.

9.7 Developing the DFD hierarchy

Figure 9.1 shows the relationship between levels in a set of DFDs.

9.7.1 Process IDs at the different levels

The hierarchy system for DFDs gives specific rules for the way IDs are allocated to the processes at the different levels of the hierarchy.

At the Context level, there is only one process box and it needs no ID.

2. Hierarchical development.

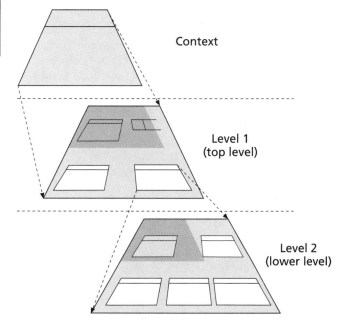

Figure 9.1 Hierarchical development.

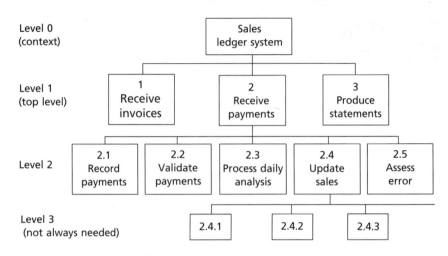

Figure 9.2   IDs in the DFD hierarchy.

At Level 1, all the processes have IDs of one number (an integer), as shown in the video store Level 1 DFD. At lower levels, process IDs are a little more complicated.

A Level 2 DFD is an expansion of a particular Level 1 process. The IDs of the Level 2 processes are related to their Level 1 process by having IDs that begin with the ID number for the Level 1 process. In the video store Level 2 DFD for **Deal with loan**, all the processes begin with the integer 1, because they represent the decomposition of process 1 at Level 1. The ID of each Level 2 process for **Deal with loan** is made up of a 1 followed by a dot and a unique integer within the Level 2 DFD. So **Handle loan details** has an ID of 1.1, and **Find video** has the ID 1.2.

At Level 3, the numbering system continues. A Level 3 DFD is an expansion of one Level 2 process with an ID of two integers separated by a dot. All the Level 3 DFD IDs begin with the level process ID number followed by another dot and another unique integer within the Level 3 DFD.

Another way of showing this is depicted in Figure 9.2.

9.8   Stopping the decomposition

You will appreciate that functional decomposition cannot continue to lower and lower levels forever. At some point it must come to an end. There is no rule for when it stops, except to say that a particular process does not need to be decomposed further when its function is simple enough and clear enough.

An asterisk at the bottom right hand corner of a process shows that the process has not been further expanded on a lower level.

9.8.1 Example 9 DFDs for a sales ledger system

A company has an IS for handling its payments for sales. This IS is called the sales ledger system and interfaces (has inputs and outputs) with the customers, the company accountant, and another IS: the order processing system. Invoices come into the IS from the order processing system and statements are sent out to the customers. Sometimes payments are returned by the customers, but sometimes they need to be told that their payment is wrong, which may result in another payment or an explanation as to why the correct payment is not forthcoming. At various times, the sales ledger system delivers an analysis of the payments received to the company accountant. Given this information, a Context level DFD can be drawn as in Figure 9.3.

It is now necessary to model the system in more detail. The invoices are received and put into the sales ledger by someone in accounts. Using this ledger, someone else in accounts produces the statements that are sent to the customers. When a payment is received, some people in accounts check the payment and update the sales ledger if it is correct. An incorrect payment results in an enquiry to the customer, which may result in some explanation or a correct payment. This section of Accounts is also responsible for sending the payment analysis report to the company accountant. A Level 1 DFD can now be generated as in Figure 9.4.

Looking at this Level 1 DFD, process 1 and process 3 look rather simple and require no further decomposition—so they are given an asterisk in their bottom right-hand corner. Process 2, however, is complicated enough to require decomposing to a lower level. We discover that, in process 2, someone in the payments section receives a customer payment and records it in the payment ledger. The payment details are passed directly to another person in the payments section whose job it is to validate the payment to see if it is correct. They also pass a summary of the day's payments to the payment analysts whose job is to produce the payment analysis report for the company accountant. Valid payments are passed to the ledger clerk who looks at the sales ledger and updates it, while erroneous payments go to the supervisor who sends a letter of enquiry to the

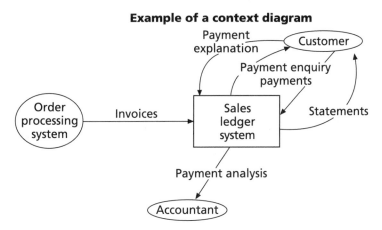

Figure 9.3 Context level diagram.

Context diagram decomposed to level 1 (top level)

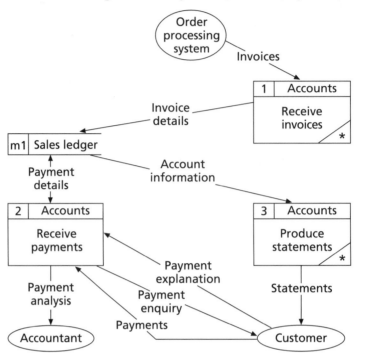

Figure 9.4 Level 1.

customer and handles any resulting explanations or excuses. From this information, a Level 2 DFD can be generated for process 2. This is shown in Figure 9.5.

The resulting sub-processes at Level 2 (2.1, 2.3, 2.4, and 2.5) seem simple
enough not too require further decomposition, but perhaps process 2.2 (validate
payments) is still too complicated and might require a Level 3. We shall not,
however, progress further with this.

9.9 Developing data flow diagrams

There are several activities needed for the successful production of the DFDs for
an IS.

- Identify the key data flows. These will result from the systems analysis paper
  chase. They will include documents, computer screens, memos, and
  telephone messages.
- Identify all the external entities. These are the sinks and sources of the IS.
- Identify all the functional areas. These will be the locations at which
  processing is carried out.
- Identify all data flow paths. You need to determine where all the data
  comes from and goes to—the source and sink of every data flow.

**Level 1 Process 'Receive payment' decomposed
to Level 2 (lower level)**

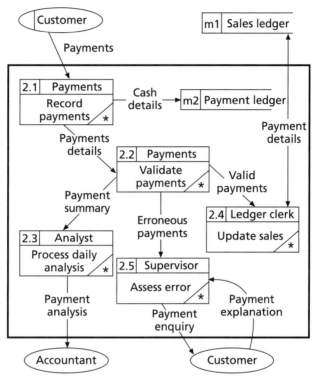

Figure 9.5 Level 2.

- Identify the system boundary. All the above will allow you to tell where your system begins and ends. Often this is obvious at an early stage of the analysis, but it will only be confirmed by knowing all of the above.
- Identify all processes. You will need to know the nature of every activity in the IS. Some processes are started (triggered) by receiving a piece of data, sometimes by a particular time and date. All these must be recorded.
- Identify all data stores. Data at rest may be in permanent or transient files. It may be historical information being held in archive. A data store may be made up of extracts from another store, or by accumulating information from several others. Some data is simply a copy of other data.
- Identify process and data store interactions. Find out what data flows from which stores to what processes and vice versa.
- Fill in the details. Add lower levels where necessary, and data flows between processes in each level.

9.10 Checking the DFD

There are several simple checks that can be carried out.

- Processes should have at least one input and one output data flow.
- Each data flow, process, store, and external entity should have unique and meaningful names.
- Exactly the same data flows should enter and leave a lower level DFD as they did in the higher level DFD. Their names must be exactly the same as well.

9.11 Tutorial 9.1

9.11.1 Wulfruna garage 1

Wulfruna garage has three main departments—servicing, parts, and accounts—which deal with customers and suppliers.

When a customer brings in a car for servicing, the service manager fills in a job sheet detailing the work to be done. He retains one copy and gives one copy to the customer. The service department requests any necessary parts from the parts department and the parts will be returned to servicing with information on any out of stock items. Customers may also buy parts directly from the parts department. The customer can order parts, and, if they are available, they can be collected immediately. Any parts that are not in stock are ordered from the suppliers. As soon as they become available, the customer is contacted so that they can be collected.

The accounts department receives information on parts supplied and goods received from the parts department, and the completed job sheets from servicing. They then produce invoices for the customers. When suppliers send their invoices, these are checked against the goods received and then they are paid. Accounts also receive payments from customers.

Draw the Context level and Level 1 DFDs to show the flow of data between departments and external entities.

9.11.2 Wulfruna garage 2

We are now going to look in more detail at the Servicing department of Wulfruna garage.

What actually happens in this department is that the service request from the customer to the servicing desk is written on a job sheet and a copy is handed to the customer, and the original is filed in the job sheets filing drawer.

A list of the parts needed is then put on a spike (the parts spike) on the desk of the parts orderer. Every so often the parts orderer empties this spike completely and fills in a parts details requisition that goes to the supplies (parts) department. If the parts are available, the supplies department delivers them; otherwise information is returned to the parts orderer on any out of stock items. This information about parts availability is passed on to the servicing desk.

Once the job is complete, the servicing desk passes the job details to the accounts department.

Draw the Level 2 DFD for Wulfruna, ensuring that it balances with your Level 1.

9.12 PG TOPIC: Questions about DFDs

There are a couple of problems with DFDs that are worth thinking about at this level.

9.12.1 Time in a DFD

One of the problems is that although a DFD appears to represent movement of data as well as activities carried out on that data, there is no representation of time in the DFDs so far discussed. In the video store example, it may seem obvious that the whole IS is triggered by a member bringing the video cassette details to the counter, but this fact is not represented in the DFD.

Structured systems analysis and design (SSADM) has ways in which time can be represented for an IS. The method for coping with time shown in the next chapter is entity life histories. This and event correspondence diagrams (not discussed here) may be worthwhile investigating before deciding if SSADM copes effectively with the time problem.

9.12.2 A decomposition too far

Also, when involved in functional decomposition, how do you know when to stop? Processes could be broken down until we are dealing with each discrete hand movement—which is what *time and motion study*[13] might want to do. However, that seems too low a level for our purposes—so when do we stop? The simplistic answer, already offered, is that the process had been decomposed enough when its function is obvious and simple enough. But it can be difficult to know when that has been reached.

We would suggest that a useful rule of thumb is to look at the data flows entering and leaving a particular process. If there is one flow going in and one going out, then it is likely that the process needs no further decomposition. If there are several flows going in and out, then that process may well need breaking down further.

But do not assume that this rule of thumb is absolute. There will be processes with only one flow in and one out, that turn out to be horribly complicated inside, and processes with, say two flows in and two out, that are really fairly simple and straightforward.

---

[13] *Time and motion study*: an investigation of the motions performed and time taken in industrial, etc., work with a view to increased proficiency and thus production. (*Chambers CD*)

Entity life histories and object oriented techniques

- [ ] The problem of time
- [ ] Entity Life Histories: One of the three views of the system
- [ ] The dynamic view of the system
- [ ] ELH notation
- [ ] Terminology
- [ ] Constructing ELHs
- [ ] Cross-checking models
- [ ] Comparing LDSs and DFDs
- [ ] PG TOPIC: Do ELHs solve the problem of time?
- [ ] Tutorial 10.1
- [ ] PG TOPIC: An object-oriented technique
- [ ] Use-case diagrams

10.1 The problem of time

As mentioned in 9.12, the techniques used so far have a problem with showing the way time affects an information system (IS). You cannot tell from a data flow diagram (DFD) what starts the whole system off, or how processes follow each other in what order over time. For the Video store example, you may guess that the system begins with a member bringing a video cassette case to the counter, but that information is not present in the DFDs. Similarly, an entity relationship diagram (ERD) tells you about the relationship between data, but it says nothing about how that data comes into existence or is deleted, nor what happens to it in the meantime.

Structured systems analysis and design (SSADM) offers two ways of trying to cope with the time problem. For processes, there is the event correspondence diagram (ECD), and for data, there is the entity life history (ELH). We shall deal in this book with ELHs.

10.2 Entity Life Histories: One of the three views of the system

IS modelling techniques are seen has having three ways of describing a system. DFDs give the *process view* and model the way information passes from one place to another. The *data view* is given by the logical data structure (LDS); this describes how the data is stored as well as the relationships between the data. The third view is the *dynamic view* of ELHs and shows how the data is brought into being, how it alters over its lifetime, and eventually is deleted. The two techniques covered so far are often seen as static views of the IS, while the dynamic view is provided by ELHs.[14]

- Process view: DFDs show how information is passed around the IS.
- Data view: LDSs show how information is related and stored.
- Dynamic view: ELHs show how information changes during its lifetime.

These three views of an IS form the stable three-legged structure intended to describe all the processes completely. This concept is shown by the diagram in Figure 10.1.

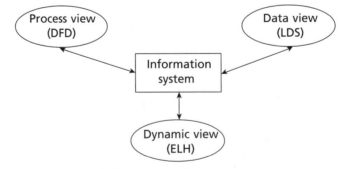

Figure 10.1 The three views of an information system.

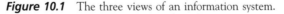

14 *Static*: Not changing, incapable of being changed.
  Dynamic: Capable of changing or being changed. (DOC)

10.3 The dynamic view of the system

This way of looking at the IS introduces the concept of ordering processes in time, so as to understand the dynamics or the behaviour of the system. An ELH models the dynamic interaction between processes and data and depicts how entities are or can be affected by events.

10.3.1 Events and effects

For entities, an *event* is a real time action that causes an update to the held data. It acts as a *trigger* for a given process, or a set of processes, to take place or kick into action. The process updates the entity but the event triggers it off. Each event causes one or more updates and each update is called an *effect*. So, a particular interaction between an entity and an event is called an effect.

10.3.2 Entity Life histories

ELHs provide an understanding of the time ordering of processes. Data has to be established before it can be used by any of the processes or before it could be modified or updated. ELM's describe in what sequence events can occur and when those events will have some effect on a particular entity. They chart the lifespan of an entity within a system from inception to completion (deletion) in terms of the events that affect it.

It becomes important here to distinguish between an entity type (or set, or class) and the individual entity. The inception and completion of an entity is the creation (or addition) of the details for a particular member of the entity type. The entity type itself—the **Customer** entity or the **Supplier** entity—does not disappear at completion; but a particular customer or supplier is added to the data at its inception and is deleted from the system at its completion. Between these times, the particular entity may have some changes made to it because of changes to address and such like. In biological terms, the entity is born, it has an interesting life, and (eventually) it dies.

The ELH documents all of the events that can affect an entity type, and applies to all the occurrences or members of the entity type. An ELH diagram is created for each entity in the LDS as a tree structure, having a fixed time sequence, running from left to right.

10.4 ELH notation

The notation for ELH diagrams is similar in representation and meaning to the diagrams used for the Jackson Structured Programming (JSP) method (Jackson, 1973). The ELH diagram uses the same structures or components as JSP:

> **ELH components**
>
> Sequence
> Selection
> Iteration

The approach assumes, as does JSP and the Structured English to be described in Chapter 11, that all processes can be modelled with these three constructs. The ELH diagram is a hierarchical structure made up of these three components.

10.4.1 Sequence

A *sequence*[15] of actions or activities is several actions following each other in time. You may like to think of a sequence as being one damn thing after another. A sequence of events is represented in an ELH as a series of boxes at the same level of the hierarchy that are read from left to right. They indicate that certain events will take place, one after the other, left to right, in that fixed sequence. An entity is created or born, it has a life span (it has an interesting life), and it is deleted when it dies. This is shown, in general terms, in Figure 10.2.

So, the entity's life sequence is represented by the series of boxes, reading from left to right. Its existence is shown as a sequence of three components: inception, followed by lifespan, and ending with completion. For the video shop example, the Video Title entity can be shown with the ELH in Figure 10.3.

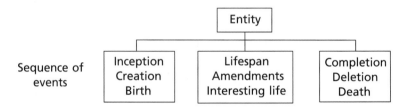

Figure 10.2　A general diagram for an ELH.

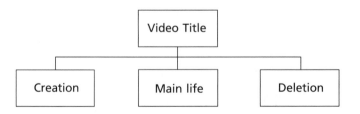

Figure 10.3　ELH for the Video Title entity.

15　　*Sequence*: order of succession; a series of things following an order. (Chambers (CD)

In the video shop example, a new video is purchased and the title is added to list of videos on offer for renting—the particular title is *created* on the system. Then this will undergo a *main life* set of events—the video is rented to various members. Eventually, the copy is so old that it has served its shelf life and it has to be withdrawn—so it is *deleted* from the system.

10.4.2 Iteration

Iteration is a repetitive series of actions that go on and on until there is no more reason for them to continue. In an ELH diagram, an iteration of events is represented by a box with an asterisk in it; this indicates that the event involved may occur once or several times (or, indeed, not occur at all). The general idea is shown in Figure 10.4.

For the video shop system, the Video Title entity can be shown as in Figure 10.5. The main life of the video title consists of an iteration or repetition of the action **loan**.

10.4.3 Selection

The final component to be considered is the *selection*. A selection of events defines a number of options or alternatives. These options are represented by a

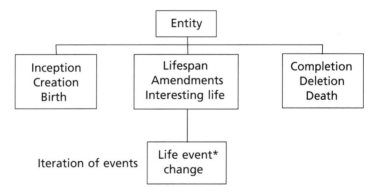

Figure 10.4   General ELH showing an iteration box.

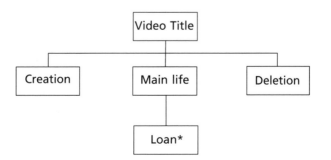

Figure 10.5   Video Title entity showing an iteration box for Loan.

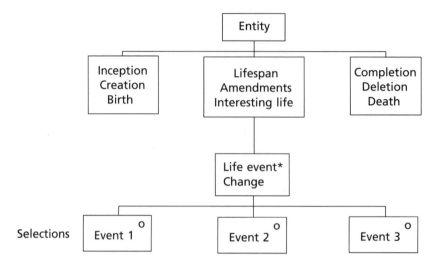

Figure 10.6   General ELH showing three selection boxes.

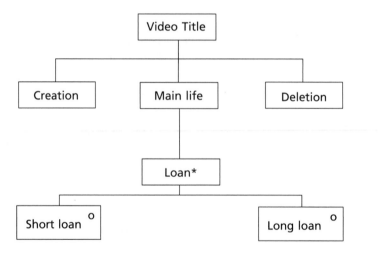

Figure 10.7   ELH for Video Title showing long and short loans as types of loan.

set of boxes at the same level, hanging from the same node (or same leg) of the structure. Each has 'o' at the corner to indicate that they are optional, mutually exclusive events, only one of which can apply or take place. This is shown, for the general case, in Figure 10.6.

For the Video Title entity, the fact that during its life it will be on loan several times, means that the iteration box for loan is needed. However, each loan may be for a standard length of time (short loan) or for twice that time (long loan), so we need two selection boxes. Any loan may be one or the other of these, but not both. The ELH for the Video Title can now be shown as in Figure 10.7.

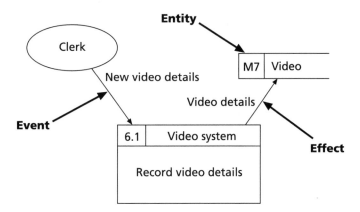

Figure 10.8   DFD of video shop IS showing entity, events, and effect.

10.5 Terminology

The ELH refers to entity types, events, and effects. The way these relate to the DFD for an IS is shown for part of the DFD for the video shop example in Figure 10.8.

Events have an effect on the entity; they change its state. Within the lives of different entities, there are similarities. Some event is needed to create an entity; bringing about its creation or birth. Another event will make a final valid change so as to delete it (the death event). Also, during its life, there will be zero, one, or more events that bring about some change to the entity. The video shop example (Figure 10.8) shows the creation of an entry in the video data store for a new video title. The primary event occurs when the clerk puts details of a new video into process 1.1 (record video details). The effect of what process 1.1 does is to add the new video title to the video data store, which represents the Video Title entity.

10.6 Constructing ELHs

So, four steps can be identified in the production of an ELH.

- Step 1: identify creation/birth events;
- Step 2: identify deletion/death events;
- Step 3: identify all main life events;
- Step 4: draw ELH.

The creation or birth event is placed on the left of the tree-like structure. The death or deletion event is placed on the far right, and the main life events are placed in the middle. These events are inserted in the correct sequence across the structure. To do this, you need to study the DFDs.

The simple example in Table 10.1 illustrates the effects of some events on the Member entity in the Video Shop system:

Table 10.1 Events and effects for the video shop Member entity.

Event	Effect
New Member	Create
Rent video	Change
Return video	Change
Renew video	Change
Member leaves	Delete

10.7 Cross-checking models

The three models we have described are intimately related to each other. So they can be used to check each other's correctness. This cross-checking is illustrated in Figure 10.9.

So ELHs need to be cross-checked with the relevant DFDs and LDS. Also, the DFDs should be compared with the LDS for the system, which we shall consider in some detail.

10.8 Comparing LDSs and DFDs

There is a strong relationship between DFDs and LDSs. As we know, the LDS of the system should reflect the relationships between the entities in the system and their data. During systems analysis, these relationships indicate the relationships that exist in the current physical system. Later, we shall see how the LDS shows how data is related in the proposed system.

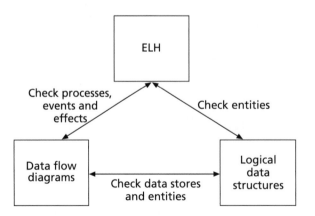

Figure 10.9 Cross-checking ELHs, DFDs, and LDSs.

On the other hand, the DFDs of the current physical system illustrate how data moves about the system in data flows, how the data is held in data stores, and how it is acted upon in processes. We said that an entity in an LDS is an object in the world that is of interest to the IS under consideration. If it is of interest, then the IS is likely to be holding information about an entity. Additionally, DFDs show the data stores within which this data is held. So, data about LDS entities are held in the data stores of the DFDs. This is just one of the relationships that hold between LDSs and DFDs; there are several others.

Data stores in the DFDs hold data, and therefore are linked to one or more LDS entities. In the final system, we should not expect to see that one entity is represented in more than one data store. However, in the current physical, paper-based system, it may well be that the data for one entity is held more than once in several manual data stores. This is the duplication of data that we shall be trying to reduce in our new system. Another exception occurs when we consider transient data stores. These do not represent permanently stored data, and do not refer to a system entity.

Also, in the DFD, a data flow consists of many data items. Clearly, each and every data item should belong to an entity.

All this means that the two types of diagram, the two views of the system represented by the DFDs and the LDS, can be compared to see that they are in agreement about the system that has been analysed. An entity on the LDS must have (at least) one data store to hold its data, and each data store in the DFD must represent at least one entity on the LDS. A data store that does not represent an entity will show up a mistake in the analysis. Either the entity has been missed on the LDS, or the data store holds data with which we are not really interested.

Relationships between DFDs and LDSs

- LDSs will or do reflect the structure of stored data.
- A DFD shows data moving about the system and being stored in data stores.
- Each data store could represent one or more entities.
- An entity may not appear in more than one data store.
- A data flow consists of data items. Each and every data item should belong to an entity.

Exceptions

- In the current system, data is often duplicated, so it is possible for an entity to appear in more than one data store. Also, link entities (cross-references) will contain the key fields from the original entities they link together.
- Transient data stores do not represent stored data and will therefore not appear in the LDS.

EXAMPLE 10

1. Draw the DFD describing the following (it describes a small area of a banking system):

   'the manager instructs a clerk to open a new customer account by providing the customer name and an account number.

   The clerk keeps the record of the customer details and opens a new bank account.

   All the transactions on the accounts are subsequently recorded by the clerks who can check the balance for a customer and deposit and withdraw money from the account'

2. Draw the LDS for the above example assuming each customer can simultaneously hold many different accounts.
3. Do the cross-checking between the DFD and the LDS on how they are related for the above example.

(The example is taken from Ashworth and Goodland, 1990)

The DFD for this partial system is shown in Figure 10.10, and the logical data structure in Figure 10.11.

Figure 10.12 shows the relationships between the two views of the banking system. The dotted lines link components on the LDS with their relevant components on the DFD. The dotted lines are labelled and explained below.

A: The **Customer** entity and **Customer/Account** link entity relate to the **Customer Detail** data store.
B: The **Bank Account** entity and the **Customer/Account** link entities relate to the **Bank Accounts** data store.
C: The **Transaction** entity relates to the **Bank Accounts** data store.
D: The **Bank Account** entity is described by a set of data items.
E: The **Balance** data item in the **Bank Account** entity relates to the **Balance** data item on the data flow.

10.9 PG TOPIC: Do ELHs solve the problem of time?

At the start of this chapter, we raised the spectre of the problem of time: the fact that DFDs and ERDs do not attempt to say when things take place, or that one event precedes another. ELHs are the way SSADM tries to cope with time, but does this technique cope well enough?

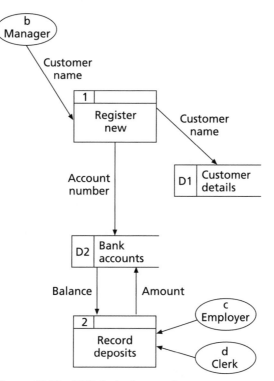

Figure 10.10 DFD for bank example.

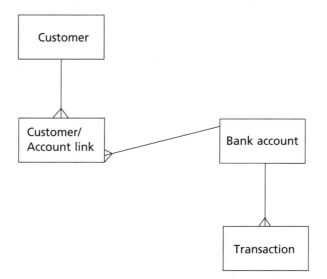

Figure 10.11 The LDS for the bank example.

Perhaps ELHs do something for the life of an entity through time; but do they solve the problem set out in Section 6.11 of how an entity may be one thing at one time and something else later on. Furthermore, the problem alluded to in Section 9.11, about time and DFDs is not really addressed at all by ELHs.

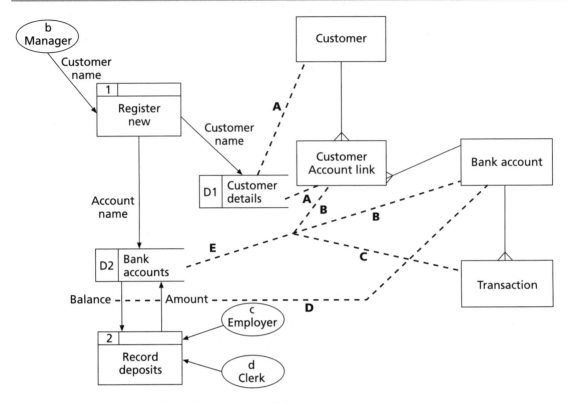

Figure 10.12   Comparing the bank IS DFD with its ERD.

It could be that, in a DFD, it does not matter what happens before what. It is true that, when an IS is underway, it is difficult to tell—and perhaps it is silly to try and tell—what started the procedure off. Yet, there seems to be some sense, from the point of view of the user and IS people, in saying that an order-processing IS begins with an order and ends with the payment of the bill. The people involved will not ignore time.

More research can be carried out by the advanced student into SSADM to see if time is represented by some other parts of the methodology such as the event correspondence diagnosis (ECDs) referred to above. Also, there will be other ways of depicting procedures in ISs and systems in general that do have ways of managing the problem of time. Students might look at the Yourdon methods, object-oriented methods, and state-transition diagrams.

10.10 Tutorial 10.1

Produce an ELH diagram for the member (customer) entity in the video shop. Use the following facts:

- a member joins and finally leaves;
- during membership, a member's address or name may be altered.

The UML allows you to represent multiple views of a system using a variety of graphical diagrams, such as use-case diagrams, class diagrams, state diagrams, sequence diagrams, and collaboration diagrams.

Hoffer et al. 1999

Language … any manner of expressing thought or feeling; an artificial system of signs and symbols, with rules for forming intelligible communications, for use in, for example, a computer.

Chambers CD

Use case A complete sequence of related actions initiated by an actor; it represents a specific way of using the system.

Actor An external entity that interacts with the system (similar to an external entity in data flow diagramming).

Use-case diagrams A diagram that depicts the use cases and actors for a system.

Hoffer et al. 1999

10.11 PG TOPIC: An object-oriented technique

It should be obvious by now that this book is dealing principally with the structured information system development (ISD) techniques that are in common use in the UK. However, the student may well be aware of another important movement for the development of ISs: *object-oriented (OO) techniques*. It seems that these methods are not in very widespread application in the UK, but more so in the USA. There is not enough space in a book like this to do justice to OO, but we shall look here at one of the techniques for capturing the functional requirements of an IS: *use-case diagrams*. This technique is part of the *unified modelling language (UML)* (see box left).

You may question whether this collection of methods constitutes a real *language*. However, it does seem to be covered by one of the definitions of a language in Chambers CD dictionary. Certainly, the set of techniques aims to cover all the necessary facilities for systems analysis and design, and to communicate what is discovered to the developers and users.

10.12 Use-case diagrams

In the object-oriented world, these diagrams cope with much the same subject as DFDs in SSADM. See box left for some definitions.

We have met the term *actor* before—in the discussion of soft systems methodology (SSM) in Chapter 5. There, actors were defined as the people who carry out the changes (transformations) to information in the system—not quite the same as here. However, you may feel that these drawings have the

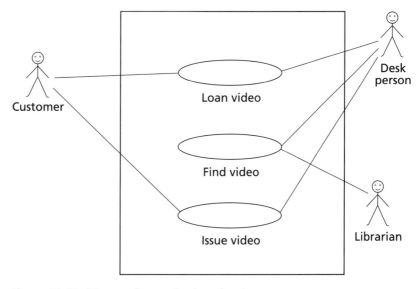

Figure 10.13 Use-case diagram for the video shop.

appearance of SSM rich pictures and have a similar function to DFDs. In use cases, the actors are just like the external entities in DFDs—they interact with the system.

Figure 10.13 shows a use-case diagram for part of the video shop IS.

Compare this with the DFDs we have drawn for the video shop IS. Sometimes, in the use-case diagram, the actors for a process seem like external entities, and sometimes they are more like DFD locations.

Process descriptions

11.1 The need to describe processes

The description of processes appears twice in this book. Later on, we shall be describing processes for the purpose of specifying computer programs. This first time deals with the description of processes that are taking place in the current physical information system (IS).

The activities taking place in the IS may well be quite complicated. We are not particularly concerned with the difficulty of carrying out some task, but the complexities involved in making decisions in the IS. Many processes need to ask questions about the state of affairs (conditions) and carry out actions depending upon the answers to those questions.

For instance, in the video store, the **Find video** process has to answer the fairly simple question of 'Does the video cassette appear in the library?' If it does then the customer will be given the cassette on loan; if not, the customer may be offered an alternative. Other processes will have more questions to answer, and the actions carried out may depend on the answers to all those questions, and the combinations of answers to those questions.

In the video library, the cassette may or may not be there, and there may or may not be one out on loan. If the video is there, then the member can be issued with it. If the video is not there and there is one out on loan, the member may be offered the chance to reserve the video. Also, if the video is not in the library and there is not one out on loan, the video has presumably been lost or stolen, so the member will have to be content with some other choice. The two questions posed (video present? video on loan?) actually produce four combinations. These are: video present and video on loan; video present and video not on loan; video not present and video on loan; plus video not present and video not on loan.

The number of combinations of conditions (questions) that needs to be coped with in a system process increases dramatically with the number of questions to answer. Therefore, in order to describe fully the current physical IS, there needs to be a method of writing down these questions and answers.

11.2 Problems with ordinary English

A common way of describing processes is to write down what goes on in ordinary English prose. For standard descriptions this may be acceptable. However, when there are many questions to note, and when the actions carried out are dependant upon the various combinations of those questions, this method soon becomes inadequate.

Also, the way ordinary English is used and understood varies substantially between writers and readers. Often, ordinary language can be long-winded, obscure, incomplete, and ambiguous. Such language may be vague, and most people are unable to use it clearly and unambiguously.

English speakers often use terms that seem clear, but are open to different interpretations and confusions. For instance, when we say that some activity is sanctioned after a person has achieved some age, we may use the phrase 'When she is over 21 . . .'. We do not mean by this 'When her age is 22 or over . . .', yet, in a computer program, writing 'If AGE greater than 21' could mean just that. What the English speaker actually means by the phrase is 'When she is over 20 . . .'; that is 21 or over. Ordinary prose might also use loose terminology, such as 'normally' and 'usually'. The systems analyst needs to be much more precise with descriptions.

Also, there are many people who use English that other people will find difficult to understand. Sometimes, this is to try to show how clever they are, and sometimes it is done deliberately to obscure the truth. The box has two examples that exemplify some of these problems:

> 1. A habitable, frame supporting, tension structure
> 2. A hexiform, rotatable surface, compression unit

The reader may find it amusing to try to work out what two common items are being referred to.

Although English is probably by far the most commonly used way of describing what is going on in an IS, for complicated decision making processes, other more precise methods have been developed. Decision making processes involve *conditions*, which are the criteria used to decide what to do (the questions); and *actions*, which are the activities to be carried out depending on the answers to the conditions. We shall describe three of these: *decision tables*, *decision trees*, and *Structured English*.

11.3 Decision tables

These are a commonly used method of handling conditions and actions in a neat way. The form of the method is a table with conditions taking up the top rows and actions being shown in the bottom rows.

> **Decision table** A table that indicates actions to be taken under various conditions, the decision being the selection between the alternative actions.
>
> *DOC*

For the sake of simplicity, we shall be dealing with so-called non-extended decision tables, where the answer to each question or the result of each condition is either a **yes** or a **no**. This binary method is fairly easy to understand, but extended decision tables may well result in smaller solutions.

As with the next two methods of description, for a decision table you need to work out all the conditions that need to be answered. In the video library there were only two: **video present?** and **video on loan?** Furthermore, all the actions that need to be carried out need to be discovered. In the video library, the member is either issued with the video, asked to reserve the one out on loan, or asked to select another video. In decision table form, this becomes:

	Rules			
	1	2	3	4
Conditions				
Video present?	Y	Y	N	N
Video on loan?	Y	N	Y	N
Actions				
Issue video	X	X	—	—
Offer to reserve	—	—	X	—
New selection	—	—	—	X

Given that **Y** is yes and **N** is no, and that **X** means carry out the action while the dash (–) means do not, this should be fairly easy to read, even with little knowledge of how to build it. There is an oddity with it. You will notice that if the video is present, the table suggests that you ask the question about the video being on loan. According to what we said before, if the video is present in the library, it doesn't matter whether some other copy is out on loan; the member is issued with the version we have found. The question of being on loan only has relevance when the video has not been found in the library. This little problem can be coped with neatly, as explained below.

We have said that the rows of the table are for the conditions and actions. The columns of the table are called the *rules*. For two conditions, we need to deal with four rules. For only one condition, there can only be two rules, one for the

Pitfall

Complex conditions
Beware of combining conditions together with **AND** and **OR**. This is good advice for process descriptions and in programming. You may well be tempted to try the condition: **Video not present AND not on loan**—do not! This seems fine when the answer is YES, but what does NO mean?

If your logic is up to it, you may be able to deduce that the **N** answer to this question means: **Video present OR on loan**. The **Y** answer covers the final column in the table safely, but the **N** answer means the same as all the other 3 columns put together; and you will still have to separate these out with more logic.

Use the KISS strategy: Keep It Simple, Stupid—Use only elementary conditions without **ANDs** and **ORs**. Believe us, complex **AND** conditions are nasty, but **OR** conditions are much worse— especially when you have to work out what their negatives mean.

yes and one for the **no**. You may like to consider how many rules there would be for three conditions—3 sets of Ys and Ns. The answer is 8.

The condition part (known in the trade as the *condition stub*) of a table with 3 conditions, and, therefore, 8 rules would look like this:

C1	Y	Y	Y	Y	N	N	N	N
C2	Y	Y	N	N	Y	Y	N	N
C3	Y	N	Y	N	Y	N	Y	N
Actions								

The number of actions in the *action stub* has no bearing on the number of rules. You should persuade yourself that every possible combination of **yes** and **no** has been catered for in this table. We now have 2 rules for 1 condition, 4 rules for 2 conditions, and 8 rules for 3 conditions. Fans of the geometric series may like to guess at how this series continues. Luckily, there is a formula for calculating the number of rules in a table with a given number of conditions. If the number of rules is **r**, and the number of conditions is **c**, then the following is true:

$$r = 2^c$$

That is, the number of rules equals 2 raised to the power of the number of conditions, or 2 multiplied by itself **c** times. So for 4 conditions there are $(2 \times 2 \times 2 \times 2)$ 16 rules, for 5 there are 32 rules, and for 6 conditions there are 64 rules. So, as we said, the rules increase very rapidly as the conditions increase by a small amount. This formula is useful to learn because using it tells you that you have the right number of rules for a given number of conditions and have not missed one or added one too many.

11.3.1 Worked example

Read the text in the box below and find all the conditions and all the actions. Then calculate how many rules there must be to cope with all the combinations of those conditions. Set all this out in a decision table.

Procedure narrative

- You are the greenkeeper at your local golf club. If the grass is too long, you cut it. If there are weeds then you apply weedkiller. Otherwise, you can sit in a deckchair and enjoy the weather.

The following box shows what you should have discovered:

> ## Greenkeeper
>
> - <u>Underlined conditions</u>
> - **ACTIONS IN BOLD**
> - You are the greenkeeper at your local golf club. <u>If the grass is too long,</u> **you cut it.** <u>If there are weeds then</u> **you apply weedkiller.** <u>Otherwise,</u> **you can sit in a deckchair and enjoy the weather.**

There are, therefore, two conditions and three actions. The **otherwise** condition (and it *is* a condition) will be catered for without making it another line in the condition stub. Two conditions means (using $r = 2^c$) there are 2^2 or 4 rules. These can be shown in the decision table as follows:

> ## Greenkeeper decision table
>
> - There are 2 conditions (therefore 4 rules); and
> - 3 actions.
>
> | | | | | |
> |---|---|---|---|---|
> | Grass too long? | Y | Y | N | N |
> | Weeds showing? | Y | N | Y | N |
> | | | | | |
> | Cut grass | X | X | – | – |
> | Apply weedkiller | X | – | X | – |
> | Sit down | – | – | – | X |

Persuade yourself that the **otherwise** condition in the text is accommodated by the last rule—the grass is NOT too long **and** the weeds are NOT present.

11.3.2 Building decision tables

The worked example only had two conditions, but if there are many conditions to consider, it can be difficult to ensure that all the combinations of those conditions have been catered for in the table. There is a straightforward way of building the condition stub so that all the possible rules are included. The first thing is to work out how many rules there are going to be.

Be careful, by the way, that you are using *elementary* or *simple conditions*. It can be tempting in programming as well as describing processes to dive into using *complex conditions*. Complex conditions are usually spotted because they include *connectors* such as **and** and **or**. An example of a complex condition is 'If the driver is young **and** drives a sports car'. This contains at least two simple conditions: 'If the driver is young' (yes or no); and 'If the driver drives a sports car' (yes or no). Believe us, using complex conditions is a dangerous business, particularly when the connector is **or**, as in 'If the driver is young **or** drives a sports car'. This may seem to make some sense when the answer to the condition is **yes**. However, when the answer is **no**, the condition becomes: 'If it is not the case that the driver is young **or** drives a sports car', confusion (and incorrect descriptions and programs) very often ensue.[16]

Once the complete set of simple conditions has been found and the number of necessary rules has been calculated, the condition stub can be built-up. For three conditions, the number of rules is 8 ($r = 2^3 = 2 \times 2 \times 2 = 8$), the first line of the condition stub is constructed by dividing the number of rules by 2 and putting down that number of **Y**s, in this case 4 of them, followed by the same number of **N**s, as follows:

Condition 1   Y Y Y Y N N N N

The next line requires the number of rules to be divided by 4, and that number of **Y**s followed by that number of **N**s, for as long as is needed to fill out the line. In this case, 8 divided by 4 is 2, so condition 2 has 2 **Y**s, 2 **N**s, 2**Y**s, and 2 **N**s, as in:

Condition 1   Y Y Y Y N N N N
Condition 2   Y Y N N Y Y N N

The last line of a condition stub is always an alternating sequence of **Y** and **N**. In fact, this is a special case of applying the same procedure as for the first two conditions. Divide the total number of rules now by 8 and put down that many (that is **1**) **Y**s and **N**s until the end of the line. In the case of three conditions, the whole condition stub now looks like this:

Condition 1   Y Y Y Y N N N N
Condition 2   Y Y N N Y Y N N
Condition 3   Y N Y N Y N Y N

For four conditions, following the same procedures, the first line is 16 divided by 2. That is, 8 **Y**s and 8 **N**s. The entire stub for 8 conditions is:

Condition 1   Y Y Y Y Y Y Y Y Y N N N N N N N N
Condition 2   Y Y Y Y N N N N Y Y Y Y N N N N
Condition 3   Y Y N N Y Y N N Y Y N N Y Y N N
Condition 4   Y N Y N Y N Y N Y N Y N Y N Y N

Building your decision table in this way should ensure that all the possible combinations of conditions will have been catered for.

11.3.3 Rationalizing decision tables

Large decision tables can be difficult to read and understand. So it is generally desirable to reduce them so that they are as concise as possible. This is carried out by *rationalizing* or *optimizing* the table. Decision tables can be simply reduced where two rules (columns) only differ by one condition. This means that the actions for the two rules must be *exactly* the same. Also, the Ys and Ns for all

---

[16]   This complex negative condition means, logically, that the driver is not young **and** not a sports car driver.

the conditions in the two rules must be exactly the same, except for only one condition, which must therefore have a Y and an N. As an example, we repeat the decision table for the video shop example set out above:

	Rules			
	1	2	3	4
Conditions				
Video present?	Y	Y	N	N
Video on loan?	Y	N	Y	N
Actions				
Issue video	X	X	—	—
Offer to reserve	—	—	X	—
New selection	—	—	—	X

Examine the first two rules. The actions are exactly the same (they both result in **Issue video**, and both do not **Offer to reserve** or to give a **New selection**). The condition stub for these rules only differs by one condition—the **Video on loan?** has a **Y** and an **N**, while the **Video present?** rules are both **Y**. This means that if the video is present, the video is issued whether or not the video is on loan. So these two rules may be combined into one rule. To do this we need another symbol for the condition stub, that symbol is a dash, which means we do not care what the answer to the condition question is. It covers both the Y and the N, as in the following:

	Rules		
	1	2	3
Conditions			
Video present?	Y	N	N
Video on loan?	—	Y	N
Actions			
Issue video	X	—	—
Offer to reserve	—	X	—
New selection	—	—	X

So four columns have become 3, but the table means exactly the same thing. In this way, large tables may be reduced to quite small ones. See the box for a more general example.

Rationalizing decision tables

- Can reduce large DTs by combining columns (rules), for example:

IF A?	Y	Y	...	N	N
IF B?	Y	Y	...	Y	N
IF C?	Y	N	...	N	—
DO THIS	X	X	...	N	—
DO THAT	–	–	...	X	X
DO OTHER	X	X	...	X	X

- Can become:

IF A?	Y	...	N
IF B?	Y	...	–
IF C?	–	...	N
DO THIS	X	...	–
DO THAT	–	...	X
DO OTHER	X	...	X

- The dash in a condition (–)
  - means Y and/or N
  - don't care/doesn't matter

- **<u>BUT ACTIONS MUST BE IDENTICAL</u>**

11.4 Decision trees

Decision tree A binary tree where every non-terminal node represents a decision. Depending upon the decision taken at such a node, control passes to the left or right subtree of the node.

DOC

This is another popular way of showing conditions and actions. The official definition (left) is rather daunting.

However, they are in fact very easy to understand. The decision tree for the video store library procedure is shown on the following page.

Most people seem to be able to understand such a diagram immediately. If the answer to the first condition (**Video present?**) is true, the yes path (**Y**) is taken. The second condition is then looked at (**Video on loan?**), and the **Y** or **N** path taken to the particular action (in both cases: **Issue video**). The **N** path for the first condition leads to the second question being asked again and the **Y** or **N** path being taken from that to the relevant actions (**Offer to reserve** for **Y**, or **New selection** for **N**).

Decision trees are generally constructed and read from left to right; but, as suggested by the definition, this could be done the other way. Trees might also be shown going from top to bottom or bottom to top. They are particularly useful where the **sequence** of conditions is important. It could be that once the answer to the first question is known, the second question is different for the **N** than for

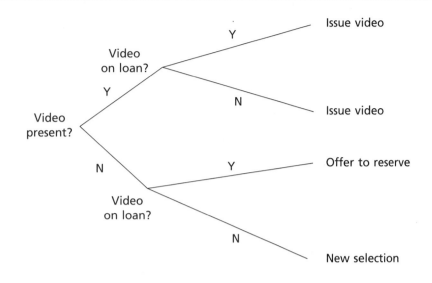

the **Y**. In a decision table, all conditions are present all the time, but in a decision tree, different conditions might be brought into play depending on the answers to previous ones.

Decision trees may be optimized in much the same way as decision tables. An optimized version of the video shop decisions would look like this:

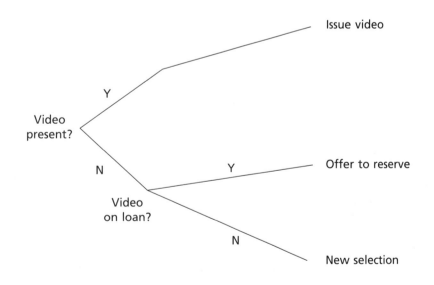

Note that all that has happened is that the second condition has not been asked if the answer to the first question is **Y**. The route goes straight to the action **Issue video**; there is no point asking if the video is out on loan if one is actually present. The two identical actions for the full version have been collapsed into one, just as the two rules with identical actions were collapsed into one in the decision table answer. This was possible because both the **Y** and the **N** route led to the same action: the two routes only differed in the answer to the final condition.

The tree for the greenkeeper looks like this:

Decision tree example the greenkeeper

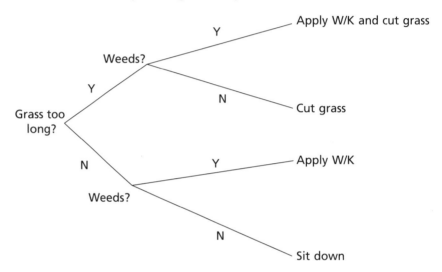

No optimization is possible here, because none of the actions is the same as any other.

These two examples deal with only two conditions. One more condition makes the tree twice as long and a little wider, and for many conditions, the decision tree can become very large indeed. Look at the worked example.

Worked example—the drinker

When you go to the bar, you may want 4 cokes, you may want 4 pints, and you may want 4 whiskies, or nothing.

After 4 pints you will be drunk, after 4 whiskies very drunk.

Produce an optimized decision table and a decision tree to show how drunk you might be (sober, drunk, very drunk, paralytic) depending on the decisions you make.

A skeleton decision table with just the stubs looks like this:

Conditions and actions

4 cokes?
4 pints?
4 whiskies?

Sober
Drunk
Very drunk
Paralytic

There are three conditions, so there will be $(r = 2^3 = 2 \times 2 \times 2) = 8$ rules. Set them out using the method laid out above. Thus:

Build the rules
Add the action Xs

4 cokes?	Y	Y	Y	Y	N	N	N	N
4 pints?	Y	Y	N	N	Y	Y	N	N
4 whiskies?	Y	N	Y	N	Y	N	Y	N
Sober	—	—	—	X	—	—	—	X
Drunk	—	X	—	—	—	X	—	—
Very drunk	—	—	X	—	—	—	X	—
Paralytic	X	—	—	—	X	—	—	—

How to optimize?

To optimize, look for identical actions in rules that only differ by one condition. In this case this is true for the following pairs of rules: 1 and 5, 2 and 6, 3 and 7, 4 and 8. The optimized version can now be drawn as:

Optimized version

4 cokes?	—	—	—	—
4 pints?	Y	Y	N	N
4 whiskies	Y	N	Y	N
Sober	—	—	—	X
Drunk	—	X	—	—
Very drunk	—	—	X	—
Paralytic	X	—	—	—

(So cokes have no effect on whether drunk or sober)

Drinker example – optimized

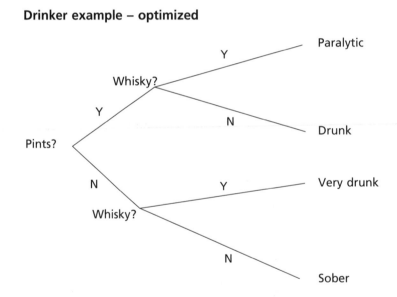

The decision tree answer would look like this:

11.5 Structured English

We have discussed some of the problems associated with the use of the natural English language in describing processes. However, there is a type of English that can be used in this area: this is *Structured English*. The box shows some more examples of how ordinary language may confuse the analyst and user.

Structured English

- WHY USE STRUCTURED ENGLISH?

Eg. 1)   Add A to B unless A is less than B, in which case subtract A from B.

Eg. 2)   Calculate total A plus B. Notwithstanding the last sentence, in the circumstance where B is greater than A the result should be the difference between A and B.

– These have the same meaning but the narrative style obscures the similarity and the sense.

The next box puts the same meaning into a much more formal and structured way of writing.

How about:

```
IF A IS LESS THAN B
        THEN SUBTRACT A FROM B
ELSE (A IS NOT LESS THAN B)
        ADD A TO B
END IF
```

You should be able to persuade yourself that all the three ways of writing describe the same thing. However, the final version seems much simpler and straightforward and less open to uncertainty. The last example is written in Structured English, and this section will teach you the basics of describing processes using it.

Structured English also has its uses as an intermediate step between the program design and its code. In that area it is often known as *Pseudocode* or *Pseudolanguage*. But here we are concerned with its use in describing a procedure that already exists in a company's IS.

The box shows the sort of terms (the *reserved words*[17] and such like) that can be used in Structured English.

---

[17]   *Reserved word*: A word that has a specific rule in the context in which it occurs, and therefore cannot be used for other purposes. (DOC)

Reserved words etc.

- IF, ELSE, END, END IF, DO WHILE, END DO, THEN.
- **IF … ELSE … END IF** statements aligned.
- Use **END IF, END DO** to delimit the blocks.
- Use **imperative** VERBS e.g. ADD, SUBTRACT, CALCULATE etc.
- A little extra English is possible, to make it more palatable.

In other forms of this technique, other reserved words are possible, including a CASE verb for selections. We are discussing the very basic necessities here.

11.5.1 The structure of Structured English

This way of writing is called structured because statements are built using the three structures of structured programming: These are *sequence*, *selection*, and *iteration*.

Sequence

When one action follows another, they are said to be carried out in sequence. Any set of actions following each other without any intervening conditions is in such a sequence. In the greenkeeper example, cutting the grass might be followed by applying the weedkiller, or vice versa.

Selection

The selection is the way decisions are taken. When a condition has two possible outcomes, the language takes one path or the other: it *selects* the next thing to do. In the video shop example, the video is either in the shop or it isn't. If it is, then the video may be rented; if not, some other decisions and actions are taken.

Iteration

It is often the case that a particular set of conditions and actions needs to be carried out several times. This repetition is called *iteration*. Generally the processes are performed until a particular condition has been met or until a particular condition is still true. In the video shop, although left unspoken so far, all of the decisions and actions to do with renting videos are carried out until the shop closes or there are no more customers.

The box reiterates these structures, and the following text describes them in a little more detail.

> ### Structure
>
> - Sequence:
>   - One or more actions which take place one after the other without interruption.
> - Selection:
>   - A series of alternative conditions, from which one condition is selected.
> - Iteration:
>   - Where a series of actions are repeated until some condition is reached.

11.5.2 Sequence structure

This is a set of actions with no conditions intervening:

- DO THIS;
- DO THAT;
- DO THE OTHER.

The actions THIS, THAT, and THE OTHER will be carried out in that order.

11.5.3 Selection structure

Used when the following statements are to be carried out ONLY if a particular condition is true. It is set out in this way:

```
IF condition A
        Action 1
ELSE
        Action 2
        Action 3
END IF
```

This means that if condition A is true, then action 1 will be carried out. However, if condition A is false, then actions 2 and 3 will be performed in that sequence. The statements, exemplified here by actions, could also be conditions.

Note the formal structure of the selection. It begins with an **IF** for the condition (which should be a simple, non-complex one). The **ELSE** reserved word is used to denote the false path for the condition; and the whole thing is finished off with the reserved words **END IF**. All selections in Structured English should take this form: IF, ELSE, END IF. We shall deal with the formal way of laying it out a little later.

11.5.4 Iteration structure

As we have said, iteration is a repetition of some statements while some condition is in force—while it is still true. The Structured English being described here uses **DO WHILE** as the reserved words to denote this. (This is sometimes known as a *DO loop*.) In general terms the iteration looks like this:

```
DO WHILE condition A
        Action 4
        Action 5
        Action 6
END DO
```

This means that **while** condition A is **true**, actions 4, 5, and 6 will be carried out. When condition A is false, the processing moves to the **END DO** (which terminates the iteration) and carries on from there. Again, the statements within the DO loop could also be conditions.

11.5.5 Laying out the text

The way the Structured English text is set out is important: it helps in reading it and ensuring that everything has been catered for. The use of capital letters for reserved words has already been mentioned, but the way the text looks on the page is significant too. This is mainly carried out by the use of indentation. Look at the example above for the **IF** statement, and note how the action following the **IF** is indented. The **ELSE** afterwards is set out in the same column as the **IF** to which it refers; and its actions are indented the same amount as the first action. Finally, the **END IF** is again in the same column as the **IF** and the **ELSE**.

Often, there will be **IF** statements within other **IF** statements—this is very useful in Structured English. An **IF** that is dependent upon another **IF** is indented with respect to that **IF**—this is called *nesting*, or *nested IFs*. The **ELSE** and the **END IF** for a particular **IF** are always set out with the same indent as their respective **IF**. The box shows this clearly.

```
                      Nesting IF Blocks
IF
•
•          IF
•          •
•          •          IF
•          •          ELSE
•          •          END IF
•          ELSE
•          •          IF
•          •          ELSE
•          •          END IF
•          END IF
•
ELSE
•
END IF
```

See that two or more **END IFs** often follow each other. The indentation makes clear to which **IF** and **ELSE** a particular **END IF** belongs.

11.5.6 Worked Example

We return to the golf greenkeeper example we used for the table and tree. The conditions have been underlined, and the actions made bold.

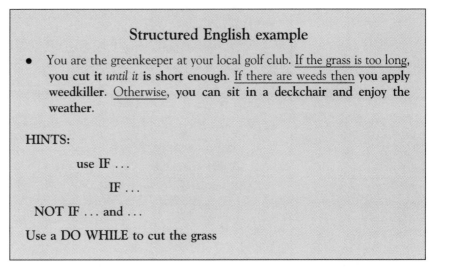

Structured English example

- You are the greenkeeper at your local golf club. <u>If the grass is too long</u>, **you cut it** *until it* **is short enough.** <u>If there are weeds then</u> **you apply weedkiller.** <u>Otherwise</u>, **you can sit in a deckchair and enjoy the weather.**

HINTS:

use **IF** . . .

IF . . .

NOT IF . . . and . . .

Use a **DO WHILE** to cut the grass

The hint in the box suggests the use of two **IFs** one following the other. This is the standard way of dealing with multiple conditions. In this example there are two simple conditions: one asks if the grass is too long, and the other asks if the weeds are showing. Once again, avoid the temptation to combine these into one complex or compound condition with an **AND**—keep them simple and separate. With three conditions there will be three indented **IFs** beginning the description; but here, we only need two.

Look at the answer box. The first **IF** asks if the grass is too long and the second if the weeds are present. Writing the second immediately after the first allows you to cope easily with the facts that the grass is too long AND the weeds are present without actually using the word **AND**. The **DO WHILE** that follows will be carried out only when **both** those **IFs** are true. This part of the text refers directly to the decision table column for **Y** and **Y** for the two conditions.

Because of the **DO WHILE**, the action **cut grass** will be performed as long as the grass is too long—the **END DO** terminates those actions that are carried out while the grass is too long. In our example world, the action **apply weedkiller** is carried out once and does not need a **DO WHILE**. It is set out in sequence following the **DO WHILE**, and finishes off those actions that are needed when both **IFs** are true (because the grass is too long and the weeds are present).

The first **ELSE** refers to the **IF** for weeds present and is therefore indented the same amount as that **IF**. The **ELSE** denotes the negative of that **IF**, and what follows the **ELSE** are those actions and conditions that need to be considered when the grass is too long but the weeds are **NOT** present. It refers to the table

```
IF grass too long
   IF weeds present
      DO WHILE grass long
         Cut grass
      END DO
      Apply weedkiller
   ELSE (no weeds, grass long)
      DO WHILE grass long
         Cut grass
      END DO
   END IF
ELSE (grass OK)
   IF weeds present
         Apply weedkiller
   ELSE (no weeds & grass OK)
         Sit down
   END IF
END IF
```

column for **Y** and **N**: **Y** for grass too long and **N** for the weeds. The words in the round brackets or parentheses are for any comments, which is allowable in Structured English to clarify what is going on.

What happens here is that the grass is cut but the weedkiller is not applied. So the **DO WHILE** for cutting the grass while it is too long is placed here. The **IF** for **weeds present** (dependent on the truth of the **IF** for **grass too long**) has now been completely dealt with. It is therefore terminated with an **END IF** indented at the same level.

With only the two conditions, we have also dealt with all the processes that take place when the grass is too long. The negative of that condition now needs to be coped with, so an **ELSE** for the first **IF** is needed at the same indentation. Note the comment in parentheses for **grass OK**. We are now looking at the two decision table columns which are **N** for **grass too long**.

When the first **IF** is false, we need to ascertain the truth of the second **IF** again; so the **ELSE** is followed by **IF weeds present**. When this is true, we are looking at the table column for **N** and **Y**; and we need to apply the weedkiller— but NOT cut the grass because it is not too long. That's all we need to do if the grass is OK and the weeds are present, but we need to provide for the final set of values for the two conditions. This is simply done by the use of the **ELSE** indented at the same level as its preceding **IF**. As the comment suggests, this denotes that the grass is OK and the weeds are NOT present. The original text says that when

Pitfall

Programmers, or putative programmers, often see Structured English as the place to demonstrate their prowess in coding. But a Structured English process description is not the same as a computer program; one prescribes to a computer what it is to do, and the other describes what is going on—so care must be taken by programmers.

this is the case ('otherwise'—corresponding to the **N** and **N** of the decision table) the greenkeeper can sit down and enjoy the weather.

The **IF** and **ELSE** for the weeds when the grass is OK needs to be terminated with an **END IF** indented at the same level. The Structured English description is almost complete, but we need to terminate the very first **IF** with an **END IF** set at the indentation for the first set of **IF** and **ELSE**. People often forget this; don't be one of them.

11.5.7 Worked example

The full Structured English answer to the drinking example may seem rather long. We repeat the text here followed by that answer.

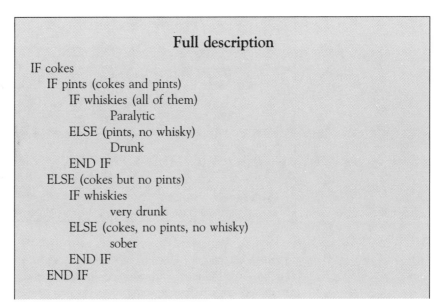

> ### Example
>
> When you go to the bar, you may want 4 cokes, you may want 4 pints, and you may want 4 whiskies, or nothing.
>
> After 4 pints you will be drunk, after 4 whiskies very drunk.
>
> Produce an optimized decision table and a decision tree to show how drunk you might be (sober, drunk, very drunk, paralytic) depending on the decisions you make.

In the answer, note how the three conditions are dealt with by three **IFs** starting off the description. In a full answer this means that the whole thing is finished off by three **END IFs** following each other at the end.

> ### Full description
>
> ```
> IF cokes
>     IF pints (cokes and pints)
>         IF whiskies (all of them)
>                 Paralytic
>         ELSE (pints, no whisky)
>                 Drunk
>         END IF
>     ELSE (cokes but no pints)
>         IF whiskies
>                 very drunk
>         ELSE (cokes, no pints, no whisky)
>                 sober
>         END IF
>     END IF
> ```

```
ELSE (no cokes)
    IF pints
        IF whiskies (no cokes, pints, whiskies)
                Paralytic
        ELSE (no cokes, pints, no whisky)
                Drunk
        END IF
    ELSE (no cokes and no pints)
        IF whiskies (whisky only)
                very drunk
        ELSE (no cokes, no pints, no whisky)
                sober
        END IF
    END IF
END IF
```

We know from the table and tree that this example can be optimized; this is also true of the Structured English. It so happens that the optimized version is simply the full version missing out the conditions about the cokes because they have no effect on the state of drunkenness. It looks like this:

Optimized

```
IF pints
    IF whiskies
            Paralytic
    ELSE (no whisky)
            Drunk
    END IF
ELSE (no pints)
    IF whiskies
            very drunk
    ELSE (no pints no whisky)
            sober
    END IF
END IF
```

(No mention of cokes)

In other examples, life may not be so easy. As with the decision tree, optimization can be more difficult with the equivalent decision table, even though the Structured English may be somewhat easier to read for some people.

11.5.8 Video example

To round things off, the optimized Structured English description for our video shop process would look like this, properly laid out:

```
IF video present
        Issue video
ELSE (video not present)
        IF video on loan
                Offer to reserve
        ELSE (Video not present AND not on loan)
                New selection
        END IF
END IF
```

Note how this mirrors perfectly the optimized decision table for this procedure, repeated below.

		Rules	
	1	2	3
Conditions			
Video present?	Y	N	N
Video on loan?	—	Y	N
Actions			
Issue video	X	—	—
Offer to reserve	—	X	—
New selection	—	—	X

11.6 PG TOPIC: Which technique to use?

With three process description techniques to choose from, there ought to be some way of choosing which to use, rather than simple, blind prejudice. The decision will be based upon the particular use to which the process description is to be put, and the type of process that is to be described.

Decision tables (when optimized) tend to be very compact compared to decision trees, but trees may be considered to be easier to draw up.

Decision trees are useful where the order of the conditions is important. If you need to decide whether or not the video is present in the library before asking whether or not a copy is out on a loan (which seems sensible), then a decision tree shows this more clearly than the equivalent table. However, it may be that the decision table condition stub could be drawn up on the understanding that the first condition that appears is considered first, and so on.

On the other hand, it seems easier to optimize a decision tree than it does a table. The conditions are easier to manipulate in a table when looking for ways of reducing the description. With a complicated tree, you may be lucky in seeing that some conditions may be ignored, but changing the drawing so as to find such irrelevant conditions can be very difficult.

Perhaps decision trees are more easily understood by non-computing people than tables or Structured English.

Writing in Structured English tends to appeal to those who write computer programs because they fancy that the two procedures are similar; whereas trees and tables are not like other things that programmers understand. The danger here is that programmers often tend to think that they are writing a program rather than describing a process. The two activities are conceptually different and serve different functions. Furthermore, decision tables have actually been used in writing computer programs or parts of them. The National Computing Centre's programming language called Filetab (later, FTL6) specified all its processing by means of decision tables. Also, there were several packages on the market that would turn decision tables into program code by means of a decision table preprocessor.

Hoffer *et al.* (1999, pp. 330–332) have a section on this question. The research they quote suggests, *inter alia*, that decision trees are better than tables for portraying simple problems and making decisions, while tables are better for complex logic, compactness, and ease of manipulation. Structured English is better than decision tables and trees for transforming conditions and actions into sequence, but worse for checking consistency and completeness.

11.7 Tutorial 11.1

11.7.1 Video store

At the video store counter, the member of staff needs to ensure that a borrower is a member and that the member's age is not too young for the class of video being issued.

Generate process descriptions in the form of a decision table, a decision tree, and Structured English for this procedure.

Use the following clues:

- Member?
- Adult video class?
- Age too young?
- Enroll new member.
- Refuse to issue video.
- Issue video

11.7.2 A mail order book club

Members of this book club, whose orders exceed £100 in a year and have a good payment record are given priority treatment.

Where a membership has been over 10 years, the member will get priority treatment if the order exceeds £100 per annum, in which case the club is prepared to overlook a poor payment record.

If the order sent by a member of more than 10 year's standing does not exceed £100 in a year, and if the member has a good payment record, then priority treatment is also given.

In all other circumstances, the order is given normal treatment.

1   Draw a decision table to represent the above information.
2   Draw a decision tree to represent the above information.
3   Write a Structured English description to represent the above information.

Logicalization

12.1 Physical and logical systems models

In order to introduce the practice of *logicalization*, we need to consider the differences between the physical models of an information system (IS) and its logical models. However, this is not an easy thing to do. The main problem is that the terms 'logic' and 'logical' are used in very different ways within information systems development (ISD). For instance, we have already met the idea of logical data structures (LDSs). The word 'logical' in LDS differentiates it, as we pointed out, from the physical expression of the data structure on computer discs, or, perhaps, in manual filing cabinets. Also, the 'logic' of a system refers to the decisions and actions taken during the processing of the system. We have dealt with these aspects in the chapters on process descriptions. There are also logic gates in the electronic components of a computer—which we shall not be dealing with at all.

The reference to logic in the term 'logicalization' is more like that in LDSs than in processing and computer logic. It refers to the way ISs can be modelled in ways that exclude their physical implementations. The data flow diagrams (DFDs) we have described so far are for the physical implementation of the current IS— the *current physical system DFDs*, which model the entire system as it is physically running, now. They refer to forms on paper, filing cabinets, in trays and spikes, and also the physical locations of where processes take place. As we said, one of the reasons for producing DFDs of the current physical system is that it is often the major requirement of a new, required, or proposed IS that it carries out the same procedures as the current system—no doubt with improvements and with its errors mitigated or eliminated.

12.2 The current logical DFD

However, the systems analyst or system designer cannot assume that the new system will be implemented in the same way as the old one. That would be to limit the activities of the proposed system to that of the old, and we wish to be free to make whatever improvements seem best for the company. Therefore, we produce a version of the IS DFD that is not limited to the physical attributes of the current one. This is called a logical model—the *current logical DFD* of the IS.

A physical data flow model is a diagrammatic description of the IS as it physically exists in the real world. It represents things as they are, complete with those constraints imposed by organizational, political, and technical factors. For instance, in the current physical system there may well be two processes for carrying out what might, logically be considered to be only one. Perhaps the current system has one person to write out the details of an invoice, and another person to calculate the total cost of that invoice. This situation could have come about for the simple reason that the company had someone who could write invoices clearly but was incapable of carrying out calculations accurately. So the two processes were instituted. However, if we ignore this physical fact of the current system, we may consider that filling in the details of an invoice and working out the arithmetic associated with it is, logically, one process not two.

Looking at the IS in this way produces a more *abstract* view of the system. It helps us to understand the underlying functionality of the current system, identify problems with the IS, and may help to establish its boundaries more firmly. It will also act as the basis for the new IS.

12.3 The process of logicalization

In order to carry out the process of logicalization, we need to perform three main tasks:

- rationalize the data stores;
- rationalize the processes at the lowest level;
- consider the data that is passed along the data flows.

We shall describe ten separate steps that should be carried out in logicalization. The order in which we describe these procedures is not all that important, but the order we give would not be a bad one to follow.

In showing these steps, we shall depict the original physical DFD (or rather part of it) on top, and the logicalized DFD underneath. For reason of clarity, we may omit some data flow name—you should not. The examples are taken from an IS about a car rental company. We discuss first, two ways of rationalizing data stores.

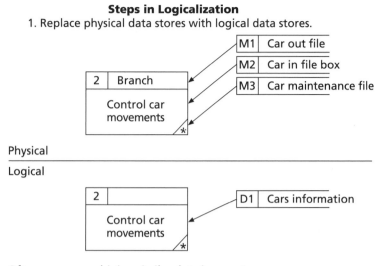

Steps in Logicalization
1. Replace physical data stores with logical data stores.

Often means combining similar data in one store

Figure 12.1 Physical and logical data stores.

12.3.1 Replace physical data stores with logical data stores

Figure 12.1 illustrates this step.

What has happened here is that the systems analyst, in conjunction with the user, has decided that the three separate physical data stores are, logically, the same data store. Perhaps there were good reasons why the original system should hold the three files rather than having just one. However, logically, they could have been held as one. In fact, they all hold data for one entity, CAR.

So the three manual stores are replaced by one logical store. The new store is given an ID starting with D to denote that it is not a physical, manual store, but a logical, database data store.

Also note, for later, that the name of the data store has been altered, and the location in the process has been deleted.

It might also be that one physical data store should logically be represented by more than one store. It may be the case that the company's IS holds several data stores in one filing cabinet, such as the details of customers and the details of their orders. In entity relationship diagram (ERD) terms, they are holding the details for two entities (CUSTOMER and ORDER) in the same store. The analyst and user may well decide that logically, they should be held as two data stores.

12.3.2 Worked example

Let us take the physical DFD in Figure 12.2 and logicalize it.

In this example, there are three stores that appear to be holding the same, or very similar, information. They are all storing details of accounts. The truth that, physically, the company has decided that they need to hold the old, new, and special accounts in different places should not blind us to the fact that logically

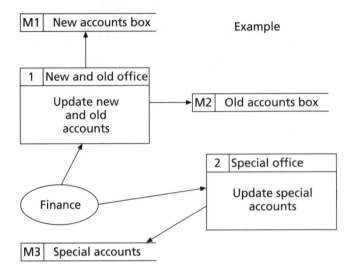

Figure 12.2 Example for logicalizing data stores.

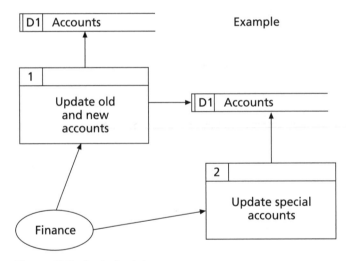

Figure 12.3 Logicalized data stores.

they all represent the entity ACCOUNT. In the logical DFD, seen in Figure 12.3, they are shown as one data store.

In the logical view the three manual data stores become one logical data store with the ID of D1.

12.3.3 Remove transient data stores

Transient or temporary data stores are a good example of physical characteristics of the current IS that exist simply because of the physical constraints of the real world. Remember that our standard examples of such data stores were the spike and the in tray. They exist because the company has decided that it is physically

2. Remove transient data stores.

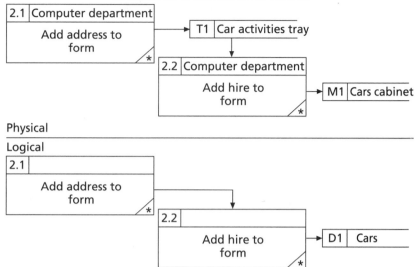

Figure 12.4 Removing a transient data store.

impractical to pass each piece of paper from one place to another as soon as they have been dealt with. Instead, the pieces of paper are stored—for the time being—until someone empties the tray or the spike and transfers the information to wherever it needs to go next.

In the logical view, the tray and the spike are obvious physical manifestations of the current system; and they are simply deleted. Figure 12.4 illustrates this step.

In Figure 12.4, information is passed from process 2.1 to process 2.2 by way of the car activities tray, designated with an ID of T1 to show it is a transient store. In the logical view, T1 has been removed. It is important to note that the connection between the two processes is not lost, and a data flow still exists between them. It is, however, no longer broken by the transient data store.

12.3.4 Remove processes that cannot be automated

Now we come to several steps about rationalizing processes. As we pointed out above, one of the reasons for logicalization is that it can help to discover the precise borders of the IS with which we should be concerned. This is because, during logicalization, we may well find procedures that we and the users can agree we could not imagine that we could automate or computerize. Figure 12.5 shows an example of this.

Process 3.1 contains the procedure of manually checking a car that we agree with the IS user we are not going to attempt to computerize. The logical view initially adds an external entity that will carry-out that procedure. Making it an external entity puts the process outside the boundary of the IS with which we are concerned.

Unfortunately, it is not enough simply to delete such a process. This is at least because we still need to know that something is going on that concerns the logical IS, even if what happens in the process is nothing that we could computerize. So

3. Remove/separate out processes that cannot be automated and/or require subjective decisions or will remain subjective.

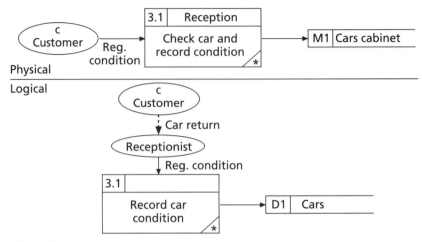

Figure 12.5 A process that cannot be automated.

the example in Figure 12.5 appears to have turned the process into an external entity, but the flow of information from the customer into the system still exists. It is shown as a dotted data flow arrow, because it now flows from an entity to an entity.[18]

There are other considerations too. In the physical DFD, process 3.1 also records the condition of the car in the data store M1. Just deleting the process would mean losing the data flow to this data store as well. But this is not necessarily something that will not be automated, and we cannot ignore it. Given that we have broken one of our rules for data flows, perhaps we could break it again, and allow a data flow from the new Receptionist external entity directly to the data store. But, in no IS can external entities—a human being, a company, or another system—have direct access to a data store. I cannot, by sheer force of will put data onto a computer disc; and I should not be allowed to walk into another person's office and put details directly into their filing cabinet. There must be a procedure—a process—within the IS to allow me to do this. So, in this case, we cannot break the data flow rule of always having a process on one end, or the other, or both. There must be a process—left behind so to speak—with a suitable name, to cope with the physical need to access a data store, either in order to put information into it, or for taking data from it. The data store M1 becomes D1 in the logical DFD.

12.3.5 Example 12

The example of the physical DFD in Figure 12.6 shows another process that cannot be automated.

---

[18] Note that this breaks one of our firm rules for physical data flows that they must begin or end with a process. In logical DFDs this particular exception is allowed.

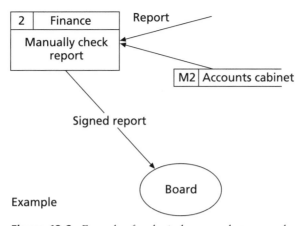

Figure 12.6   Example of a physical process that cannot be automated.

Here, a report comes into process 2, where it is manually checked and signed. After that, the report is passed to the external entity representing the board of directors. As well as the fact that manually checking a report and signing it has been agreed not to be something that we will computerize, it is important to note that the data store M2 (Accounts) needs to be looked at in order to carry out the manual checking. Accessing this data store must be allowed within our logical IS, even if the act of manually checking the report cannot.

The first step is to create an external entity, within which the manual checking will be done. A reasonable name for this external entity would seem to be found in the location of the process that we are removing (Finance). But the Finance entity still needs to access the data store **Accounts**, so we need to leave behind a process to allow it that access. The final logical DFD is shown in Figure 12.7.

Note that the signed report data flow now goes from Finance to Board as a dotted line, and the original report must go to the created Finance external entity—as long as there was a process to send it there originally. If not, another process will have to be drawn to allow that too. Also, the data store M1 becomes D1, and a suitable name has been found for the process that has been left behind.

12.3.6 Remove processes that only reorganize data

In manual ISs, some people have jobs in which they take some paper and shuffle it. They do not change what is on the paper, but it is originally in an order that is not useful, and their task is to sort it into an order that will be more so. Sadly, in today's computer age such exciting tasks are less common because computers have many ways of reorganizing data. The outcome of this is that, in the logical view of the IS, such data reorganization can be removed. In Figure 12.8, the process to sort the car activities, from their disordered state into a file of ordered (or sorted) car activities is, simply, deleted.

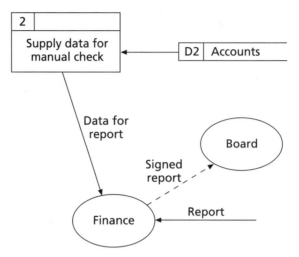

Figure 12.7   Logical view of the process that cannot be automated.

4. Remove processes which only reorganise data (e.g. sorting).

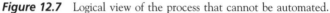
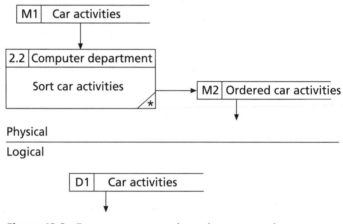

Figure 12.8   Removing processes that only reorganize data.

12.3.7 Removing processes that do not alter data

The above is an example of a more general set of processes that may be removed from the logical DFD. There are other procedures that are carried out in manual systems that do not alter data. An example of this is an operation to extract information from a data store. Again, this is very common in all ISs, manual and computerized, but the logical view of the system need not concern itself with such procedures. They are removed.

In Figure 12.9, someone has the job of riffling through the customer filing cabinet and producing details for a particular customer. Many such processes may be needed in the current manual IS, and even in the final design of the new system, but we shall deal with them later. They are not required for the current logical DFD.

**5. Remove processes which do not alter data
(*ad hoc* retrievals without creation or update).**

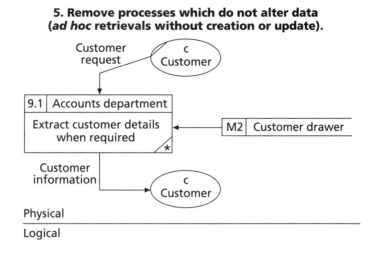

Physical
_____

Logical

Figure 12.9   Removing a process that does not alter data.

12.3.8 Combine processes joined by only one data flow

In long-standing manual ISs there are many reasons for the way procedures are carried out. Some of these reasons are because of traditions within the company, some are because of the size of jobs or the physical layout of the offices, and some are because of the need to give the boss's nephew[19] a job. In these cases, it is often discovered that one procedure passes information directly to another procedure so that, in effect, there are two or more people collaborating to carry out one task. In the logical view, two or more such processes can be combined together to produce one logical process. The way they can be spotted is often because the current physical DFD has two processes that are joined together by only one data flow. This is a sign that the two processes involved may really be the physical implementation of one logical process—and the analyst and user may agree that, logically, they may be combined into one. Figure 12.10 shows one example of this.

Here, process 2.1 passes a partially processed form to process 2.2, which carries out some more tasks with it before storing it in the data store M1. As long as there is only the one data flow between them, they may be combined together as one process in the logical DFD.

12.3.9 Worked example

Figure 12.11 shows another such example, where process 1.2 passes details to process 1.3, and from there to move M1.

Figure 12.12 shows the logical view. The two processes have become one.

_____

[19]   The word *nepotism* comes from the Italian for nephew. *Nepotism: undue favouritism to one's relations and close friends, originally by a pope. (Chambers CD)*

6. Combine processes which are joined by a data flow only, and which form part of the same or a very similar process.

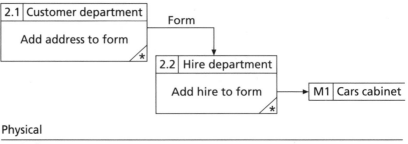

Physical
_____

Logical

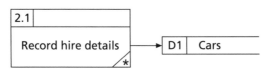

Figure 12.10   Combining two processes into one.

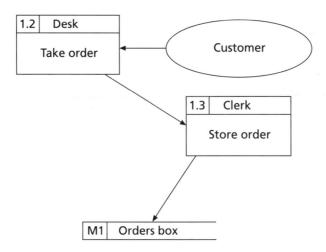

Figure 12.11   Two processes joined by only one data flow.

12.3.10   Combining processes that perform the same function

Often, company's find that they have several people doing the same job in different departments. For example, given that they often store the same data in different places, it is not surprising that they would have one person storing that data in one department and another storing the data in another. Logically, the data stores are the same, and the processes for storing the data are the same—so they may be shown as one process in the logical DFD. Figure 12.13 shows just such an example in a current physical DFD.

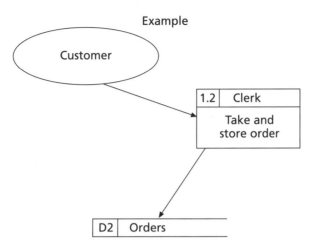

Figure 12.12 Two processes have become one.

7. Combine processes performing the same functions (appear separate only because of physical location).

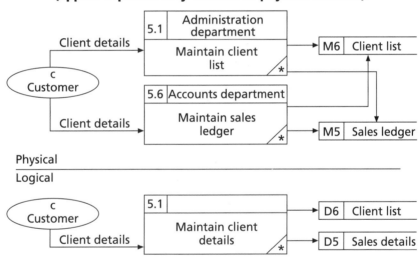

Figure 12.13 Combining two processes that are logically the same.

In this example, two people, one in the administration department and one in the accounts department, both have the job of taking details about clients and updating the data stores for the client list and the sales ledger. In the logical view the two processes become one.

12.3.11 Remove processes that exist because of tradition or politics

Sometimes two processes are found that exist as two only because of accident or tradition or office politics. They may not merely be two because of nepotism, but

8. Remove processes that exist due to tradition or politics.

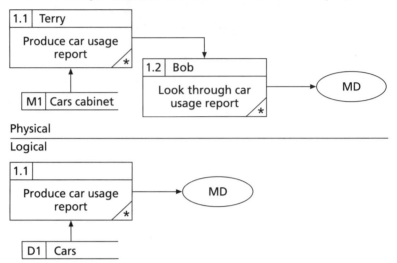

Figure 12.14   Removing a traditional process.

because the entire job was too much for one person. Whatever the reason, two physical processes that are logically only one must be shown as such on the logical DFD. Figure 12.14 shows such an example.

In this example, Bob (in process 1.2) has been given the task of 'looking through' the report before it is passed to the managing director. In the logical view, such a process may be agreed to be unnecessary and removed. (Perhaps Bob is the MD's grandson and needed a job).

12.3.12 Minimize data flow contents

In the physical DFD, a data flow may appear to be a conduit for the whole of a data store. The systems analyst may have interviewed a user and discovered that a member of staff reads the company catalogue of goods for sale in order to carry out some task—and it is this that the current physical DFD shows. However, it is very unlikely that the whole of the catalogue is going to be used for any single task. More likely, only certain aspects or parts of the catalogue will be looked at. It is important to discover the precise data that is being used and will therefore flow along the data flow joining a process to a data store. Figure 12.15 shows this as a step in logicalization.

Note that, in the physical view, the whole of the catalogue contents seems to flow along the data flow joining M9 (the product catalogue) to process 6.1 to produce the invoice. In fact, the important parts of the catalogue are likely to be only the product unit price and perhaps any discounts that may be applied. It is this detailed set of data items that is shown in the logical DFD.

12.3.13 Removing all physical references

The final task in the production of the current logical DFD is to remove the

9. Minimize data flow contents.

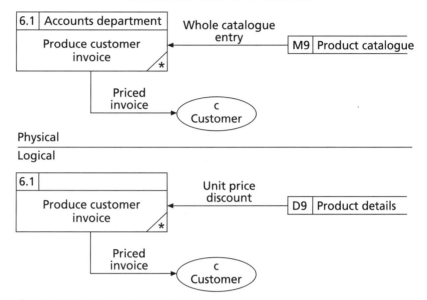

Figure 12.15 Minimizing data flow content.

vestiges of physicality by altering all references to physical objects. In the logical DFD there should be nothing about physical objects such as filing cabinets, forms, boxes, trays, spikes and such like. Less physical and more logical terms must be used for them that will not force the new IS to stick to the old ways of doing things. Also, as you will have noticed throughout our examples, the logical DFD shows no locations at all. The location where a process is carried out, or the people that carry it out, are very physical, and the logical DFD does not show them. Who knows where or by whom or by what the new IS will have its processes actioned? Figure 12.16 gives several examples.

In this example, all the physical references have been changed to logical ones. The verb 'type' has been changed to 'produce' in process 6.1. Other good verbs to use here include 'process', 'carry out', and 'perform'. Any references to the manual way information is held or passed around are also altered. So 'folder', 'page', and even 'posted' are changed. In data stores (which, of course, all have logical IDs beginning with D) names such as customer cabinet, or accounts box, or address Rolodex must be altered to be non-physical as well. In the example, the name 'product folder' is changed to be just 'product', which is acceptable. Other possibilities are 'product details', 'product information', 'product data' and such like. Also, the location 'Accounts department' is simply deleted.

12.4 The order of logicalization

As we suggested at the beginning of this chapter, the order that we have presented the steps of logicalization is not intended as a sequence where you

10. Remove all physical references.

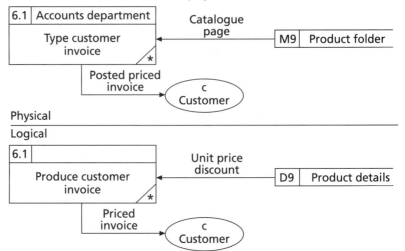

Physical
_____
Logical

Remove: type, page, folder, posted, location

Figure 12.16   Removing all physical attributes.

carry out one step followed by the next, in order. However, the succession of steps is not too far from what we would advise you to follow. Certainly, there is no point in removing all physical references, such as locations and filing cabinets, before you have reduced the DFD as much as possible by deleting and combining processes and data stores. So, this is the last step you should carry out. Also, the removal of a transient data store often exposes the fact that two processes are now joined by only one data flow. So the removal of spikes and trays is a good idea before looking at combining or deleting processes.

The order of logicalization

- combine and delete data stores;
- delete transient data stores;
- combine and delete processes;
- minimize data flow contents;
- remove all physical references.

12.5 PG TOPIC: The need to logicalize

The topic of logicalization tends to be very new for students, so a few thoughts on its necessity would seem to be in order. Perhaps its justification stems from the need to model the current physical system in the first place. If you remember, the current physical system is where the systems analyst goes for a great deal of the requirements of the proposed system. This is because, often, the task is to computerize a paper-based or partially computerized IS that is in use at the moment.

Once the current physical system is modelled, the analyst wants to model the proposed system, but the proposed system will very likely look not much like the old one. Most importantly, for computer people and users, the new system will have computers in it while the old one may not. Also, other additions and improvements will be made. The problem with the model of the old system is that it is restrictive in the way it does its work. There are real people in real offices, carrying out activities using real paper and phones and filing cabinets and all the other physical paraphernalia of bureaucracy. The new system should not be bound by these physical constraints of the old.

The steps between the model of the current physical system and the new one are therefore carried out in the more abstract, non-physical world of the logical model. Here, there are no real people and desks, no filing cabinets or spikes, and no paper. A new system can then be worked out, with all its improvements without being tied to the way the old system did it.

It is also important to understand, that the logical view of the system is not always a way of getting to a new completely computerized IS. When the new physical system design is worked out, real people and offices and such like will be back again, and there will probably still be lots of paper. The logical view steps back from the old physical constraints but should not fall into the trap of limiting itself to some view of full computerization that is just as physical and just as restricting.

12.6 Example 12.1

Logicalize this Level 1 DFD (see diagram on page 158).

The way the logicalized version was produced was as follows:

- Data store M1 was combined with M2 because they hold the same sort of data. They become D1.
- For the same reason, M3 and M4 were combined to become D2. Store M5 becomes D3.
- The transient store T1 was removed.
- Process 1 was combined with process 2 because, with T1 gone, they are now joined by only one data flow and they seem similar procedures.
- All locations were removed, as were all other physical references such as form, box, cabinet, and bill.

12.7 Tutorial 12.1

Logicalize this DFD (see diagrams on pages 158 and 159).

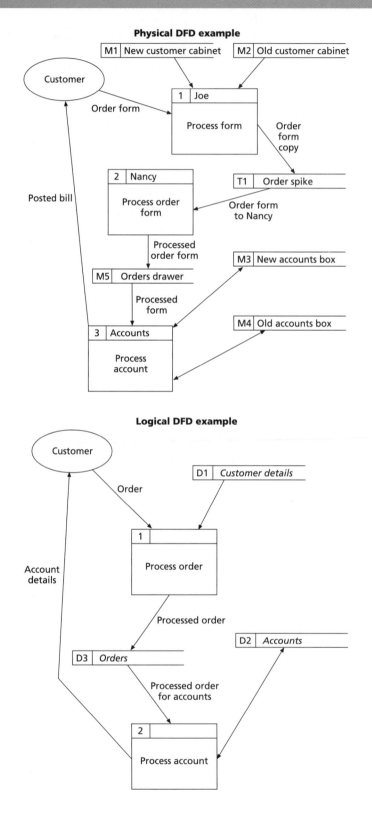

Physical DFD example

Logical DFD example

Tutorial physical DFD

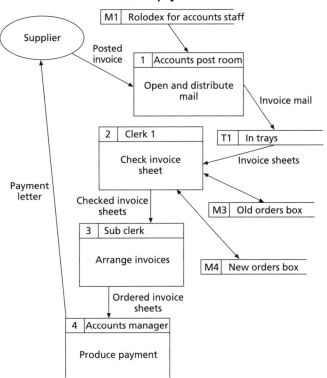

Design of information systems

- The need for design
- IS design from the systems analysis
- Specifying the proposed information system
- The desirable features of good IS design
- The design of the architecture of the system
- Business and Technical System Options (BSOs and TSOs)

13.1 The need for design

At the beginning of the book, we discussed the way any complex human endeavour requires the activity of finding out what is wanted from it—this has been covered in the preceding chapters and is called information systems (ISs) analysis. The next essential stage, we said, was to plan and design what is wanted. This, in information system development (ISD), is ISs design, or the design of IS, or systems design. Before rushing on to buying hardware and writing computer programs, modern business practice insists that any new project must be thoroughly planned and designed.

We have covered part of this already. In Chapter 12 on logicalization, we have already moved to some extent from the analysis of the current system. We have, through logicalization, moved to the current logical system, which is the abstract view of the current physical system.

Systems design is important because it allows system developers to manage several difficult aspects of a new IS:

- **Complexity:** An information system tends to be a complicated thing. Careful, checked design can help IS staff to cope with this complexity and give them a fighting chance of getting most of it correct.

- **Maintenance:** The hope is that carefully designed systems will be easier and cheaper to maintain.
- **Meeting user requirements:** Clear design will help the user to be assured that the IS that is to be built is the one that fulfills the needs of the business.
- **Control of costs:** Large computer ISs are very expensive things. A systematic design process should produce clear designs and give understandable and unambiguous routes to the new system. This allows the possibility of controlling the costs of the system, because what is to be done and why it is to be done is clear for everyone to see and follow.
- **Correctness of the system:** No computer system can be proved to be absolutely correct,[20] but a good, clear design can help the developers to produce test cases with which to put the parts of the IS to the test. The aim of this is to give the developers and users much more confidence in the new system.
- **System documentation:** The system design becomes part of the documentation of the final system. It is the blueprint of the new system and allows developers to produce the system that is needed and enhance its maintenance.

  Documentation aids communication between the IS developers and the users of the system. Program specifiers and programmers will use the design document to develop the programs for the system. It provides something that can be tested and checked by users and developers. The design will also be used to provide the patrons of the new system (those who are paying for it) with an aid to making decisions about what precisely they are willing to have developed for their money.

13.2 IS design from the systems analysis

A very important point that must never be lost sight of, is the fact that an IS is intended to be the solution of a business problem or a set of business problems. The purpose of the IS is to improve the business. IS design is concerned with the translation of the business requirements uncovered during systems analysis into possible and practical ways of meeting those needs. All the information that has been gathered during analysis feeds into the design process to produce a blueprint for the new or proposed IS.

Referring back to Chapter 3, the **requirements catalogue** contains the functions that the business has said that it needs from the new IS. The **current** logical data structure diagram is the model of the way the data is related within the current system, and the current physical and logical data flow diagrams (DFDs) and process descriptions model the way the present system works. Systems design puts all this information together to produce the **proposed** or **new** IS. The output of these deliberations will be the **proposed logical DFDs**.

---

[20]   For a discussion of this, see Chester (1998 and 1999).

In Chapter 14 we shall be learning about the technique of relational data analysis or normalization. This produces, in a systematic and fairly straightforward way, the files or tables that are going to be required for the new system. We shall also show how, together with the current logical data structure (LDS), the normalized tables can be used to produce the proposed LDS for the new system.

13.3 Specifying the proposed information system

One of the outcomes of the design process for a computerized IS will be a set of specifications for the new system. Many of the processes of the proposed DFDs will become the computer programs for the new system and will need to be specified using, in part, the process description techniques covered in Chapter 11. The new system will also need ways for the user to interact with the IS. This requires the design of inputs and outputs for the system (input–output design), incorporating the techniques of menu and dialogue design.

13.4 The desirable features of good IS design

Much can be said about what makes good design. We shall discuss the topic by considering eight of its features.

- **Functionality:** This is often considered to be of prime importance for a computer system. The **functionality** of a computer system, consists of what the system **does**—its functions. It should be patently obvious by now that the new computer system should have the functions that the business requires. It should not deliver less than what was agreed it would, and it should also not provide much more than the user has agreed to pay for.
- **Efficiency:** This is a measure of the way resources are used to produce a particular result. The fewer the resources, the more efficient is the production.

   In ISs, one of the most important and scarce resources available is time. An efficient IS is able to cope with the general turnover of information in the time that has been agreed with the user. So it will need, for instance, to process a specified number of documents per hour or day. Another time factor with which the new system will be concerned is its response time: the time it takes for the system to respond to the input of a user. There is also the run time for a particular process: the entire time to complete that particular procedure. An efficient system will meet the users' needs in the use of time in all these respects. A user of the system will quickly become disillusioned if delays are unacceptably long or agreed timescales are not achieved.
- **Flexibility:** Many organizational systems are dynamic. They are affected by internal growth and politics, staff resignations and appointments,

administrative reviews and rearrangements, external takeovers, policies and pressures, and variations in customer preference and behaviour. It is inevitable therefore that the ISs that support such organizations will need to be altered to keep up with their changes. Systems that are not built so as to be flexible and cope with change, will become progressively irrelevant.

- **Portability:** One of the ways companies need to alter their usage of computerized ISs is that they sometimes need to run those systems on different computer hardware. A system may need to be transported from one computer to another.

    The investment in an existing system must not be lost by being unable to move it to another computer platform. Therefore, IS designers must take effective measures to ensure that only minimum conversion work is required to transfer systems from one computer environment to another.

- **Security:** Computer software and data are costly and hence valuable assets to the business. So an IS must be resistant to damage and breaches of privacy and confidentiality. Systems must also be designed to meet legislative requirements for privacy (for instance the Data Protection Act).

- **Reliability:** A computer system will be expected to be reliable. There are many measures of system reliability, which we do not intend to explore thoroughly here. Part of the reliability of the software and hardware of a system may be seen as the amount of time that the system is available for use before it fails. In Chapter 3 we referred to the concept of *mean time to failure* for this. The data of the system must also be reliable in that it should be correctly recorded, without omissions, stored safely, and not corrupted or lost—either accidentally or deliberately.

- **Economy:** Here we are referring to economy of storage for data and computer programs. There should be as little duplication and redundancy of data as possible.

- **Usability:** The IS should be as easy and simple as possible to learn and use.

> **Portable**   A word applied to software that can be readily transferred to other machines.
>
> *DOC*

13.5 The design of the architecture of the system

In this book, the architecture of an IS is the way data, programs, and interfaces work together to deliver an effective system to the user. The final product of the systems design is the blueprint for this architecture, and will be used by the developers to produce the new system. Such an architectural design can be pictured as consisting of three tiers or levels.

13.5.1 Three-tier architecture

A simple representation of this architecture using DFD components is shown in Figure 13.1.

In this diagram, the leftmost component (shown as a data store) represents the design of the physical data—the database. The middle (shown as a DFD process) is the design of the program handling the data—generally an enquiry

Figure 13.1 Three-tier architecture.

or an update. The final component (shown as the DFD external entity, **user**) stands for the design of the interface between the user and the computer program—this means the design of the menus and dialogues through which the user accesses the data by means of the program.

One tier of the design interacts (shown by data flow arrows) with one other tier. This means that the user interface interacts with the program, and the program interacts with the database, but the user interface cannot directly interact with the database. The analogy with the rules of DFD modelling should be obvious, where an external entity may only interact with a data store via a process.

The point of looking at the design of the architecture in this way is that each of the three tiers may, in the final computer system, be put on a different machine. The database could be on one computer, the program on another, and the user interface on yet another. Thus, this method supports the design of client-server and distributed ISs. In practice, such a design may well mean that maintenance is easier because each tier can be operated on independently of the others.

13.6 Business and Technical System Options (BSOs and TSOs)

It is rare for only one solution to be possible for the particular organization's business problems that are under consideration. The systems analyst and designer ought to be able to provide the company with several possible answers to their requirements. These solutions are discussed here at two levels: the business level and the technical level. The possible answers to the problems for the business lead to several options for the organization to consider; these are called the *Business System Options* or *BSOs*. Each BSO may have more than one way of being implemented technically; these are called the *Technical System Options* or *TSOs* for each BSO.

13.6.1 Business System Options

The BSO for the system design is of very great importance. It governs the final objectives for the design and the completed computer IS. As mentioned above, there ought to be several BSOs for the company's management to consider, and no professional IS developer ought to provide only one option for the business needs. The differences between each BSO may be small or great. The simplest way to offer more than one solution is to offer solutions for different sets of the company's requirements.

The systems analysis will have discovered many requirements that the company might like for its new computer system. Not all of these requirements will be affordable in time or resources, and not all will be technically feasible. One BSO therefore would be for the proposed computer system to incorporate every one of the expressed requirements. Another BSO could be to try to produce a system with only the essential features, leaving out all the nice things that the company might like but that are not indispensable. Between these two extremes there are going to be various combinations of requirements, including all of the essential ones and some of the wish list ones as well. So several BSOs can fairly easily be generated.

BSOs are produced after the requirements catalogue (RC) has been completed. Each entry in the RC is considered and its entry in a particular BSO is recorded along with the reasons. Perhaps two to six options could be produced in this way. Using the RC, the project team will be able to establish a list of the *minimum* **Functional** and **Non-Functional Requirements** that all the business system options must satisfy. Look back at Chapter 3 for a discussion of these different types of requirement.

The box lists the factors that should be considered for each BSO.

BSOs—consider the following factors:

- Approximate cost of each option
- Development timescale
- Known technical constraints
- Organization of the system—types of access (online or offline) and interfaces with other systems
- Approximate data and transaction volumes
- Training requirements
- Benefits and impact on the organization

It is also possible, and perhaps less confusing to management, for the IS team to shortlist three or four BSOs, and to give their recommended solutions. However, it is unprofessional to load the dice so that in effect only the IS team's preferred solution seems to be sensible or possible. Management should be offered a real choice. Each of the shortlisted BSOs should be formally presented to the decision makers, along with their strengths and weaknesses so that a decision is made possible.

The selected BSO should be formally documented along with the reasons for its selection, and the reasons for rejecting the others. This decision is very important because it forms the basis for the whole of the rest of the project and the final system.

13.6.2 Technical System Options

Now that the BSO has been chosen, the company must agree the way that the solution is to be implemented. Once again, there are likely to be several ways that the system can be implemented; several TSOs should be provided by the ISD team.

One of the first things to do in this endeavour is to consider the constraints for the project. These can be separated into constraints external to the project and the internal constraints.

13.6.3 External constraints

There will be many restrictions from outside the project. The box lists some of these:

External Constraints

Organizational policy—There may be a policy on the use of packages, software houses, and in-house development, etc.

Costs—In an organization's IT budget, there may be maximum or minimum costs specified for any project.

Time—The required delivery date for the new system will have enormous influence on implementation planning. This could easily rule out the entire strategy.

Hardware and software platforms—Very often the developer may have to use specific languages and operating systems. Also, the hardware may have to be of a particular type or be purchased from a specific vendor or manufacturer. Even with a freer choice, standards may be set by a company's IT strategy.

Benefits—A company may set financial objectives that define a system's financial performance or savings.

13.6.4 Internal constraints

The user of the project will generate their own restrictions for a TSO. The box lists some of these:

Internal constraints

Service levels—This is a common constraint concerning the availability of system, its reliability, and its recovery time should it fail.

Performance—This is mainly concerned with response times.

Capacity—The maximum number of users, transactions, and data storage volumes.

Security—Concerns such facilities as passwords and logon procedures.

Priority—Those areas of the system that have priority and which areas are critical.

Other mandatory facilities—There can be many of these, such as the fact that the system must have email and Internet access.

Once again, several TSOs should be developed, and two or three picked out for more detailed development. The final TSO needs to be selected carefully with the full understanding and agreement of management and the IS team. It should be fully documented, along with the reasons for its selection over the others.

The box shows the main parts of the documentation for each TSO.

Technical Environment description:
This describes the hardware, software and file sizing for the system:

Hardware—Diagrams and text define the type and number of devices and networks. This will also include the needs of the environment and installation. There should also be information on reliability, maintainability and the possibilities of upgrading the hardware.
Software—Covers the details of application packages, operating systems, databases, construction software, and system utilities.
System Sizing—This describes the system's ability to store and process data.

System description:
This relates the TSO to the Requirements Specification. There should be little variation in this between TSOs, because most major decisions were taken at the BSO stage. It should make it clear what facilities are **not** being provided by the option.

Impact Analysis:
This covers the effect of the TSO on the organization:

Training Requirements Description—Covers the training needs for the TSO; who needs it and what it costs.
User Manual Requirements Description—Deals with such items as the system facilities, screens, help, security, and errors.
Testing Outline—This outlines the testing strategies that will be adopted.
Take-on Requirements Description—Covers data conversion and changeover strategy, along with their timing and costs.
Organization and staffing—Deals with all the personnel aspects of the TSO.

Cost-Benefit Analysis[21]:
Development costs—Covers the cost of hardware, software, training and others.
Operating costs—Deals with the costs of such items as labour, maintenance, and consumables.
Benefits—These are often divided into *tangible* and *intangible benefits*. Various financial performance measures are used.

---

[21]   **cost-benefit analysis** or **cost-benefit assessment:** the comparison of the cost of a particular course of action with the benefits (to be) derived from it. (Chambers CD)

Relational data analysis (normalization)

- Designing data tables
- Normal forms
- Un-normalized form
- First normal form
- Second normal form
- Third normal form
- Testing for third normal form
- Worked examples for RDA
- Rationalizing normalized tables
- Converting normalized tables to a logical data structure
- PG TOPIC: Different approaches to normalization
- Tutorial 14.1
- Tutorial 14.2

Normalization A technique for producing a set of relations with desirable properties, given the data requirements of an enterprise.

Connolly and Begg, 1999, p. 192

14.1 Designing data tables

At some stage in the design of the new information system (IS) the designer needs to plan what data will be needed as well as the way that data will be organized.

In order to produce the files for the new IS, this book uses the method developed by Codd (1970) and Date (1975), which is based on the mathematical theory of sets. It is called **relational data analysis (RDA)** or, more commonly, **normalization**.

Redundancy The provision of additional components in a system, over and above the minimum set of components to perform the functions of the system.

DOC

This definition is rather too general. For our purposes here: for *components* read *data*. Normalization is intended to minimize redundancy of the data in the system.

Relational database model A data model that represents data in the form of tables or relations.

Hoffer et al., 1999

A *table* is in more traditional parlance, a *file*. A table has *rows* and *columns*, which, more traditionally, are known as *records* and *fields* (or *attributes*).

Anomalies Errors or inconsistencies that may result when a user attempts to update a table that contains redundant data.

Hoffer et al., 1999

In fact, it is not only redundant data that can cause update anomalies. Any defective database may result in the loss of data or inconsistencies between the data.

The aim of this technique is to produce computer files (or **tables**) that are flexible and keep redundancy to a minimum.

This lack of redundancy optimizes the storage of data within the **relational database**.

Using this technique, data items are grouped together in a sensible way so as to keep storage costs to a minimum, to allow the database to be altered as easily as possible, and to allow it to be updated without causing errors in that data. In the jargon, such errors are known as **update anomalies**.

Another advantage of well-normalized tables is that they are flexible, so that, when a table needs to be altered (changed, added to, or reduced), this can happen with the minimum of effect to other tables in the database. A disadvantage may be that—as you will see—the technique tends to produce lots more tables than perhaps you might have thought necessary. This has an overhead of requiring a lot of time and resources, both human and computer, to access the data and keep it in good order.

14.2 Normal forms

The normalization technique utilizes the concept of **normal forms**.

This book will deal with Codd's (1970) original three normal forms as we think there are enough challenges in these for a book of this type. For later forms, we suggest you consult books specializing in database design, such as Date (1975) and Connolly and Begg (1999). The method described here is known as the **four-column method** of normalization (the reason for this will become obvious), which is commonly used in places where structured systems analysis and design (SSADM) is being followed.

The four columns are set out as shown in Figure 14.1.

Astute readers will notice that there are actually more than four columns here. However, the columns headed UNF, 1NF, 2NF, and 3NF are the four in question.

The three forms are introduced by the **un-normalized form** or **UNF**; drawn up from the known data items required for the new IS and discovered through systems analysis. This list of data items is refined to the first of the normalized

UNF	LEVEL	INF	2NF	3NF	TABLES

Figure 14.1 The layout for the four-column method of RDA.

<table>
<tr><td>

Normal form A
state of relation that
can be determined by
applying simple rules
regarding dependencies
to that relation.

Hoffer et al., 1999

</td></tr>
</table>

Steps in RDA

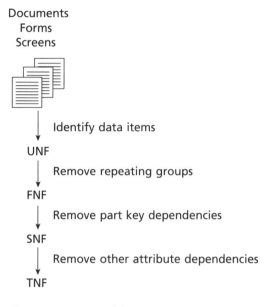

Documents
    Forms
    Screens

|
Identify data items
↓
UNF
|
Remove repeating groups
↓
FNF
|
Remove part key dependencies
↓
SNF
|
Remove other attribute dependencies
↓
TNF

Figure 14.2   Normal forms.

forms: called the **first normal form** or **FNF**. Further work produces the **second normal form** or **SNF**, and eventually the **third normal form** or **TNF**. These steps can be shown as in the flow diagram of Figure 14.2.

Figure 14.2 also gives some hints about the procedures that convert the data items from one normal form to the next. We shall consider them in turn, beginning with the very important first step of drawing up the list of data items in the un-normalized form.

14.3 Un-normalized form

This is the basis of the relational data analysis of the data items for the new system, so it is essential that it is created carefully. During systems analysis, many pieces of information will have been identified during the paper chase (see Section 3.2.3) of the current physical system. Each of these pieces may well become a data item, or attribute, of the entities in the new IS. Looking at the requirements catalogue (RC) may also suggest other data items that should also be included in the new system. Remember that the paper chase looked at forms, screens, memos, telephone calls, and other ways in which information is passed around during the operation of the IS. The requirements analysis may also have suggested that other forms and screens and such like will be needed by the new system. The data items for these must also be included in the normalization.

In the four-column method, once all the data items have been found for a particular form, for instance, they are written down one below the other in the

Key A key is an attribute or a combination of attributes which can be used to uniquely identify an entity

Goodland and Slater, 1995

Repeating group A set of two or more multivalued attributes that are logically related.

Hoffer et al., 1999

first of the four columns. Then a **key** attribute is selected from the entire list of data items.

In fact, there may be more than one attribute or group of attributes that uniquely identify an entity (more loosely, a record). You may see these referred to as *candidate keys*. The selected key for a table is often called the **primary key**. For instance, both the Employee Number and the National Insurance Number uniquely identify an employee in the Employee table; but, much more commonly, the Employee Number is selected as the primary key.

At this stage we can also look to see if the data contains **repeating groups**.

In simpler terms, it is any group of data items that is repeated several times, often in the form of a list. So the **item number**, **quantity** and **price** in a list of details in an order form, is a group of attributes that repeats.

So repeating groups are normally seen in the data sources as lists of data items. An order form often contains a list of articles to be purchased. Each such article may well be identified by an item or catalogue number and its description. It will be accompanied by the quantity of that item to be ordered as well as its price and the cost of these articles. As hinted at in our definition, the item number, description, quantity, price, and cost make up a group of data items for an order line. These order lines, each made up of its group of data items, constitute the list of items ordered—a repeating group. Such a list of an order form is shown in Figure 14.3.

The way we shall deal with such repeating groups in un-normalized form is to write the **level number** of each data item in the column next to the column for the UNF. A level number of 1 means that the data item is not part of a repeating

---

Welsh clothing
Order form

Order Number: AB1242
Order Date: 22/2/2003

Customer Number: C5432
Customer Name: Dai the Death
Customer Address: Flat 4,
Above the Undertakers,
Llareggub,
Wales

Item number	Description	Quantity	Price £	Cost £
234	Green socks	4	1.20	14.80
532	Yellow hats	5	12.00	60.00
653	Purple people	1	123.43	123.43

Cost of Order: £188.23
VAT: £ 32.94
Total Cost: £221.17

Figure 14.3 An order form.

group, while a level number of 2 shows that the data item *is* part of such a repeating group. A level number of 3 is used to show a repeating group *within* a repeating group that has already been put down as Level 2. For this basic introduction to normalization, we shall not be dealing with groups at Level 3 or greater.

From our order form, we can find a number of data items or attributes. We list them below with the repeating group indented:

> Order number
> Order date
> Customer number
> Customer name
> Customer Address
> > *Item number*
> > *Description*
> > *Quantity*
> > *Price*
> > *Cost*
> Cost of order
> VAT
> Total cost

The primary key for the order form in Figure 14.3 will be *Order number*, and the UNF column of the RDA would look like that in Figure 14.4. Traditionally, a key is underlined (we also put it in bold for emphasis).

Note that the data items for the list of items ordered are all put down as Level 2, because they constitute a repeating group. All the other attributes are Level 1.

UNF	Level	FNF	SNF	TNF	Tables
Order number	1				
Order date	1				
Customer number	1				
Customer name	1				
Customer address	1				
Item number	2				
Description	2				
Quantity	2				
Price	2				
Cost	2				
Cost of order	1				
VAT	1				
Total cost	1				

Figure 14.4 UNF and levels for the order form.

UNF	Level	FNF	SNF	TNF	Tables
Order number	1	**Order number**			
Order date	1	Customer number			
Customer number	1	Customer name			
Customer name	1	Customer address			
Customer address	1	Cost of order			
Item number	2	VAT			
Description	2	Total cost			
Quantity	2				
Price	2	**Order number**			
Cost	2	**Item number**			
Cost of order	1	Description			
VAT	1	Quantity			
Total cost	1	Price			
		Cost			

Figure 14.5   FNF for the order form.

14.4 First normal form

FNF   A relation that contains no repeating data.

Hoffer et al., 1999

The purpose of the next column is to put the data items into FNF.

Tables in FNF will have removed repeating groups. In order to carry this out, any repeating group in the UNF is put into a table of its own. This new table requires a key, which is created from a sensible key attribute (or more than one if necessary) for the repeating group combined with the original key for the UNF set of attributes. The normalization columns now look like that in Figure 14.5.

The FNF column now shows two tables. All the data items at Level 1 are in one table, and the items for the Level 2 repeating group are in another. The key for the Level 1 attributes remains as it was in UNF (*Order number*), while the key for the separate table for the repeating group is made up of the initial key and a key attribute from the repeating group. In this case, the repeating group key is (fairly obviously) *Item number*. By adding the key from the initial set of attributes to a key from the repeating group, a connection or link is made between the two tables. Without such a link, there would be no way of telling what items on order belong to which order number. Such a combination of attributes as a key, formed from the keys of other tables so that they are linked together, is called a **compound key**.

SNF   A relation . . . in first normal form and every non-key attribute is fully functionally dependent on the primary key

Hoffer et al., 1999

14.5 Second normal form

The aim of putting the attributes into SNF is to ensure that any attribute in a table is completely dependent on the whole key for that table.

This raises the knotty concept of **dependency** or **functional dependency**.

Given two attributes A
and B of a relation, B is
functionally dependent
on A if whenever any
two tuples (rows) of
the relation have the
same value for A, they
will necessarily have
the same value for B.

DOC

The DOC definition needs a little deciphering. It may help to think of an *attribute* as a *data item*, a *relation* as a *file*, and a *tuple* or *row* as a *record*. So the definition says that in a file, if two records have the same value for data item A, then they will also (and always and necessarily) have the same value for B. Another way of looking at dependency is that B is dependent on A if, when you know the value of A, there can only be one value for B. So, when you know the customer number, there can only be one customer name; consequently, the name is said to be functionally dependent on the customer number.

For our order form, the *description* and *price* of an item can be always known from simply knowing the *item number*. They are dependent on only the *item number*, only part of the compound key—they are therefore said to have a **part-key dependency**. While the *quantity* required, and the *cost* of ordering several of that item can only be known by knowing the *item number* **and** the *order number*. This is because, for a different order (and hence *order number*) a different *quantity* of that *item number* will be ordered. For these attributes we therefore need to know both attributes of the compound key. They are said to be dependent on the whole of the compound key.

Now we know about the part-key dependencies for the order form attributes, we can remove them to produce a set of tables in SNF. In a similar way to what happened for FNF, the part-key dependent attributes are removed into a table of their own. The key for this new table is simply the key field (or fields) on which they are dependent. For our present example, the normalization now looks like Figure 14.6.

This is an important place to pause and see what we have done. We now have three tables. Two tables have simple keys (*Order number* and *Item number*), while the third has a compound key made up of the two simple keys. What we

UNF	Level	FNF	SNF	TNF
Order number	1	**Order number**	**Order number**	
Order date	1	Customer number	Customer number	
Customer number	1	Customer name	Customer name	
Customer name	1	Customer address	Customer address	
Customer address	1	Cost of order	Cost of order	
Item number	2	VAT	VAT	
Description	2	Total cost	Total cost	
Quantity	2			
Price	2	**Order number**	**Order number**	
Cost	2	**Item number**	**Item number**	
Cost of order	1	Description	Quantity	
VAT	1	Quantity	Cost	
Total cost	1	Price		
		Cost		
			Item number	
			Description	
			Price	

Figure 14.6 SNF for the order form.

have done is to create a **link entity** for a many-to-many relationship between the entities *order* and *item*. We could probably have predicted that one order would consist of several items, and that a particular item could appear on more than one order. We have already learned that such a $m:n$ relationship should be broken down into two one-to-many relationships with a link entity in the middle. Indeed, the entity relationship diagram (ERD) for the system might well show just that already. As if by magic, the normalization procedure has done this for us; and what is more, it has told us what attributes belong to which entity. Perhaps you now have a better idea of the usefulness and power of relational data analysis. Now we must continue to the TNF.

14.6 Third normal form

> *TNF*   A relation ... in second normal form and no transitive relationships exist.
>
> *Hoffer et al., 1999*
>
> *Transitive dependency*   A functional dependency between two (or more) nonkey attributes.
>
> *Hoffer et al., 1999*

We have ensured that all our attributes are dependent solely on the key or keys for their tables. However, there is another type of dependency that we have yet to consider—**non-key dependencies**.

In the SNF tables, there may well be attributes that do not really depend on the key field(s) of their SNF table at all. Instead, they are dependent on some other attribute that is not contained in the key for the table. They are dependent on another non-key attribute and therefore exhibit a non-key dependency. In TNF, as in SNF, such non-key dependent attributes are removed into a table of their own. However, what happens next is very different from SNF. The key for the new table has to be the attribute on which the non-key dependent attributes are dependent. So that attribute becomes the key field of the new table, but a connection must be left between the new table and the table from which it was removed. If not, the link between the two will be lost. So the key attribute is left behind in its old table to act as the linking field—such an attribute is called a **foreign key**. In this book, a foreign key is denoted by putting an asterisk next to it in its old table.

For the order form, *customer name* and *customer address* are not dependent on the key for their SNF table, they are actually dependent on *customer number*. These attributes are removed to a new table, with *customer number* becoming the key of a new table and being left behind to be the foreign key to link the new table to the old one. So TNF looks like Figure 14.7.

Another bit of magic has taken place here. We could probably have reasoned that the relationship between the entities *order* and *customer* is a one-to-many relationship. A customer places many orders, while one order is placed by only one customer. This relationship has automatically been created by the normalization to TNF. The two entities are linked together by the foreign key of *customer number* remaining in what is, in effect, the order entity.

The final step to take, at this stage, is to give the tables that have been created through the RDA sensible names. We shall call them, from top to bottom, ORDER, CUSTOMER, ORDER LINE, and ITEM. The final normalization columns now appear as in Figure 14.8.

UNF	Level	FNF	SNF	TNF
Order number	1	**Order number**	**Order number**	**Order number**
Order date	1	Customer number	Customer number	* Customer number
Customer number	1	Customer name	Customer name	Cost of order
Customer name	1	Customer address	Customer address	VAT
Customer address	1	Cost of order	Cost of order	Total cost
Item number	2	VAT	VAT	
Description	2	Total cost	Total cost	**Customer number**
Quantity	2			Customer name
Price	2	**Order number**	**Order number**	Customer address
Cost	2	**Item number**	**Item number**	
Cost of order	1	Description	Quantity	**Order number**
VAT	1	Quantity	Cost	**Item number**
Total cost	1	Price		Quantity
		Cost		Cost
			Item number	
			Description	**Item number**
			Price	Description
				Price

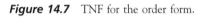

Figure 14.7   TNF for the order form.

UNF	Level	FNF	SNF	TNF	Tables
Order number	1	**Order number**	**Order number**	**Order number**	Order
Order date	1	Customer number	Customer number	* Customer number	
Customer number	1	Customer name	Customer name	Cost of order	
Customer name	1	Customer address	Customer address	VAT	
Customer address	1	Cost of order	Cost of order	Total cost	
Item number	2	VAT	VAT		
Description	2	Total cost	Total cost	**Customer number**	Customer
Quantity	2			Customer name	
Price	2	**Order number**	**Order number**	Customer address	
Cost	2	**Item number**	**Item number**		
Cost of order	1	Description	Quantity	**Order number**	Orderline
VAT	1	Quantity	Cost	**Item number**	
Total cost	1	Price		Quantity	
		Cost		Cost	
			Item number		
			Description	**Item number**	Item
			Price	Description	
				Price	

Figure 14.8   The complete RDA for the order form.

14.7 Testing for third normal form

We have completed our normalization to TNF, but we should check that the final tables are in fact in TNF.

> Test 1: For each of the non-key attributes, is there just one possible value for a given value of their key?
>
> Test 2: Do all non-key attributes depend directly upon their key?

You should be able to convince yourself that, for our Welsh Clothing example, the final tables are actually in TNF. Below, we summarize what we have done in RDA.

> ### Normalization—Summary
> - Write out data items in UNF.
> - Convert to FNF by separating repeating groups.
> - Convert to SNF by separating data items that depend on only part of their key.
> - Convert to TNF by separating any attributes not directly dependent on their key.
> - Apply TNF tests.

14.8 Worked examples for RDA

Normalization is not a simple process. Please take some time to follow the examples below and complete the tutorial that follows.

Worked example 1 concerns a medical records IS. The piece of evidence from which we shall construct our tables is the patient record card shown in Figure 14.9.

Data items or attributes for this card are put into the UNF column as in Figure 14.10. *Patient number* has been selected as the key.

Note that, for this simplest of examples, all the attributes are at Level 1—there are no repeating groups. It has been our experience that students expect always to find a repeating group for RDA. This worked example is to show that this is not always the case and to demonstrate how such a circumstance should be handled.

In this very simple case, the normal forms FNF and SNF are exactly the same as the UNF, as shown in Figure 14.11. The arrows are a short way of showing that exactly the same data items appear in one column as the next. However, in TNF, some attributes are found not to be dependent on the *patient number* key and

---

Normalization – worked example (1a)

Patient card

Patient number: 401 023 2137   Health Authority name:

Patient name: Health Authority address:

Patient address:

Doctor code: Doctor name:
Surgery Address:

Figure 14.9 The patient record card.

Normalization – worked example (1b) *UNF – write data items.*

UNF	Level	FNF	SNF	TNF	Table names
Patient number	1				
Patient name	1				
Patient address	1				
Doctor code	1				
Doctor name	1				
Surgery address	1				
Health Authority names	1				
Health Authority address	1				

Figure 14.10 UNF for the patient record card.

require other tables and foreign keys. *Doctor name, surgery address, Health Authority name*, and *Health Authority address* are all dependent only on *doctor code*. In TNF another table is created with the key of *doctor code*, and this field is left behind as a foreign key in the *patient number* table.

Furthermore, it should be obvious that *Health Authority address* is really dependent upon *Health Authority name*. So yet another table needs to be created in TNF, with *Health Authority name* as key. But *Health Authority name* must also be left behind as a foreign key in the *doctor code* table so that the link

Normalization – worked example (1c) *UNF – write data item.*					
UNF	Level	FNF	SNF	TNF	Table names
Patient number	1			**Patient number**	Patient
Patient name	1			Patient name	
Patient address	1			Patient address	
Doctor code	1			* Doctor code	
Doctor name	1	→	→		
Surgery address	1			**Doctor code**	Doctor
Health Authority name	1			**Doctor name**	
				Surgery address	
Health Authority address	1			* Health Authority name	
				Health Authority name	Health Authority
				Health Authority address	

Figure 14.11   Complete RDA for the patient record card.

between the two is preserved. In entity terms, a doctor works in one health authority, and one health authority covers several doctors.

The second worked example concerns the treatment for a particular patient. The patient's drug card is shown in Figure 14.12. This example is a simplified version of that given in Ashworth and Goodland (1990).

This is more complicated than the patient card example because there is a repeating group – the list of drugs for the patient. This is shown as a series of attributes put at Level 2 in the UNF shown in Figure 14.13. The drug code is fairly likely to be the key for the repeating group, so the compound key is **patient no** and **drug code**.

For first normal form, as shown in Figure 14.14, the repeating group is removed to another table and a suitable compound key is chosen. We chose **Patient No** and **Drug code**.

For second normal form, the attributes dependent only on *drug code* are put into their own table with *drug code* as the key. This is shown in Figure 14.15.

For third normal form, it is noted that *ward name* is actually dependent on the non-key attribute of *ward no*. So another table is created with *ward no* as key, and *ward no* is left behind as a foreign key. The final RDA is shown in Figure 14.16, in which sensible table names have also been produced.

In fact, for this example, perhaps the date and time ought to be included as part of the key for repeating group, along with the drug code. The reader should attempt another solution along these lines.

The third worked example is based on a monthly report of the sales of a salesperson for a tool company. An example of the report is shown in Figure 14.17.

		Drug Card		
Patient number: 923	**Surname:** Fawkes		**Forename:** Guido	
Ward number: 10	**Ward name:** Greenfields			

Drugs prescribed:

Date/time	Drug code	Drug name	Dosage	Length of treatment
20/5/88 12:00	CO2355P	Cortisone	2 pills × 3 day after meals	14 days
20/5/88 12:01	MO3416T	Morphine	Injection every 4 hours	5 days
25/5/88 11:00	SERA4HT	Serotonin	Injection every 8 hours	3 days
26/5/88 13:30	PE8694N	Penicillin	1 pill 3 × day	7 days

Figure 14.12 Drug card for patient.

UNF	Level	FNF	SNF	TNF	Relation (Tables)
Patient number	1				
Patient surname	1				
Patient forename	1				
Ward number	1				
Ward name	1				
Prescription – date/time	2				
Drug code	2				
Drug name	2				
Dosage	2				
Length of treatment	2				

Figure 14.13 UNF for the drug card.

The UNF for this report is shown in Figure 14.18 with *salesperson no* as the primary key. Note the repeating group – the list of customers.

The repeating group is catered for in FNF as shown in Figure 14.19. Customer number (*customer no*) has been selected as the key for the repeating group.

In SNF, those attributes dependent upon *Customer no* alone have been put in their own table with *customer no* as the key. This is shown in Figure 14.20.

Once again, it should be obvious that *warehouse loc* is really dependent upon *warehouse code*, so a new table is needed in TNF with *warehouse code* as its key, and *warehouse code* left behind as a foreign key. This is shown in Figure 14.21, where names have also been allocated to the tables.

Now, strictly, this simple RDA for the salesperson report will not do. All we have at the moment is a set of tables that tells us what sales have been made by

UNF	Level	FNF	SNF	TNF	Relation (Tables)
Patient number	1	**Patient number**			
Patient surname	1	Patient surname			
Patient forename	1	Patient forename			
Ward number	1	Ward number			
Ward name	1	Ward name			
Prescription – date/time	2				
Drug code	2	**Patient number**			
Drug name	2	**Drug code**			
Dosage	2	Prescription			
Length of treatment	2	date/time			
		Drug name			
		Dosage			
		Length of treatment			

Figure 14.14 FNF for the drug card.

UNF	Level	FNF	SNF	TNF	Relation (Tables)
Patient number	1	**Patient number**	**Patient number**		
Patient surname	1	Patient surname	Patient surname		
Patient forename	1	Patient forename	Patient forename		
Ward number	1	Ward number	Ward number		
Ward name	1	Ward name	Ward name		
Prescription – date/time	2				
Drug code	2	**Patient number**	**Patient number**		
Drug name	2	**Drug code**	**Drug code**		
Dosage	2	Prescription	Prescription		
Length of treatment	2	date/time	date/time		
		Drug name	Dosage		
		Dosage	Length of treatment		
		Length of treatment			
			Drug code		
			Drug name		

Figure 14.15 SNF for the drug card.

the salesperson to each customer. However, if this is a monthly report, then there should be a way of holding information about what has been sold to which customer for this month and every other month. With these tables that would be impossible. The reader should try to produce another solution that will cater for the months in this way. As a hint, include the **date** as part of the primary key for the UNF along with the **salesperson no**.

14.9 Rationalizing normalized tables

After RDA, the tendency is for there to be rather more tables than you may have thought there would be. Sometimes this is necessary to reduce redundancy, but

UNF	Level	FNF	SNF	TNF	Relation (Tables)
Patient number	1	**Patient number**	**Patient number**	**Patient number**	PATIENT
Patient surname	1	Patient surname	Patient surname	Patient surname	
Patient forename	1	Patient forename	Patient forename	Patient forename	
Ward number	1	Ward number	Ward number	*Ward number	
Ward name	1	Ward name	Ward name		
Prescription – date/time	2			**Ward number**	WARD
Drug code	2	**Patient number**	**Patient number**	Ward name	
Drug name	2	**Drug code**	**Drug code**		
Dosage	2	Prescription	Prescription	**Patient number**	TREAT-
Length of treatment	2	date/time	date/time	**Drug code**	MENT
		Drug name	Dosage	Prescription	
		Dosage	Length of treatment	Length of treatment	
		Length of treatment		date/time	
			Drug code	Dosage	
			Drug name		
				Drug code	DRUG
				Drug name	

Figure 14.16 Final RDA for the drug card.

Handy Tool Ltd.

Monthly Salesperson Report

Salesperson number: 1001   **Date:** May 1999
Salesperson name: Bloggs, J.
Sales area: West

Customer number	Customer name	Warehouse code	Warehouse location	Sales
101	Smith & Co.	2	Bilston	12,950.00
102	Jones & Son	4	Oxley	21,255.00
103	Smithsons	7	Parklane	10,199.80
			Total monthly sales	44,4404.80

Figure 14.17 A monthly salesperson report.

after the initial attempt at normalization there are often ways in which the resulting tables could and should be merged together. This is called *rationalizing* the tables.

Look at the final set of tables for the examples shown in Figures 14.12 and 14.16. You will see that between them are two tables for patient information. It is likely that this pair of tables actually represents just one set of data. They could

UNF	Level	FNF	SNF	TNF	Relations (Tables)
Salesperson number	1				
Salesperson name	1				
Sales area	1				
Date	1				
Total monthly sales	1				
Customer number	2				
Customer name	2				
Warehouse code	2				
Warehouse location	2				
Sales	2				

Figure 14.18 UNF for the salesperson report.

UNF	Level	FNF	SNF	TNF	Relations (Tables)
Salesperson number	1	**Salesperson number**			
Salesperson name	1	Salesperson name			
Sales area	1	Sales area			
Date	1	Date			
Total monthly sales	1	Total monthly sales			
Customer number	2				
Customer name	2	**Salesperson number**			
Warehouse code	2	**Customer number**			
Warehouse location	2	Customer name			
Sales	2	Warehouse code			
		Warehouse location			
		Sales			

Figure 14.19 FNF for the salesperson report.

therefore be merged into one table each. The rules for merging tables are as follows:

● Merge tables that share a primary key.
● Merge tables with matching Compound/Composite keys.

The two patient tables fit the first of these rules and so should be merged. The final set of tables for the combined answers to the patient card and the drug card can be shown in Figure 14.22. Note how the two foreign keys, different for each patient table have both been included in the merged table.

Of course, although we are showing here only the key fields and foreign keys, all the attributes from each of the merged pairs of tables appear in the merged tables.

UNF	Level	FNF	SNF	TNF	Relations (Tables)
Salesperson number	1	**Salesperson number**	**Salesperson number**		
Salesperson name	1	Salesperson name	Salesperson name		
Sales area	1	Sales area	Sales area		
Date	1	Date	Date		
Total monthly sales	1	Total monthly sales	Total monthly sales		
Customer number	2				
Customer name	2	**Salesperson number**	**Salesperson number**		
Warehouse code	2	**Customer number**	**Customer number**		
Warehouse location	2	Customer name	Sales		
Sales	2	Warehouse code			
		Warehouse location	**Customer number**		
		Sales	Customer name		
			Warehouse code		
			Warehouse location		

Figure 14.20   SNF for the salesperson report.

UNF	Level	FNF	SNF	TNF	Relations (Tables)
Salesperson number	1	**Salesperson number**	**Salesperson number**	**Salesperson number**	Sales-person
Salesperson name	1	Salesperson name	Salesperson name	Salesperson name	
Sales area	1	Sales area	Sales area	Sales area	
Date	1	Date	Date	Date	
Total monthly sales	1	Total monthly sales	Total monthly sales	Total monthly sales	
Customer number	2				
Customer name	2	**Salesperson number**	**Salesperson number**	**Salesperson number**	Sales service
Warehouse code	2	**Customer number**	**Customer number**	**Customer number**	
Warehouse location	2	Customer name	Sales	Sales	
Sales	2	Warehouse code			
		Warehouse location	**Customer number**	**Customer number**	Customer
		Sales	Customer name	Customer name	
			Warehouse code	* Warehouse code	
			Warehouse location		
				Warehouse code	Ware-house
				Warehouse location	

Figure 14.21   Final RDA for the salesperson report.

14.10 Converting normalized tables to a logical data structure

It is a useful exercise to turn the tables resulting from RDA into an LDS. We have learnt how to generate entity relationship diagrams from a look at the overall system, which gives us some clues about the way data is related. That is often called a *top-down* view. However, now we have worked out the detailed normalized tables for the system, we can take the *bottom-up* view as well.

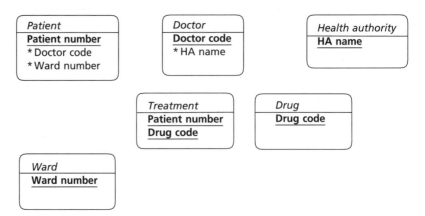

Figure 14.22 Merged set of normalised tables for patient card and drug card.

Following a simple set of rules, from the final set of tables in Figure 14.22, we can generate on ERD. The rules are as follows:

- Create an entity for each table.
- Put the primary key and any foreign keys (marked with an asterisk) in the lower part of the entity box.
- There should be an entity for every part of a compound key. If one is missing, create an entity with the missing part as a key.
- A 1 : m relationship line is put between two entities where the whole of the key of one entity appears as part of the key the the other entity.
- A 1 : m relationship is put between two entities where the primary key of one is a foreign key of the other.[1]

See Weaver *et al.* (1998) for a more detailed explanation of this procedure. The resulting ERD is shown in Figure 14.23.

The reader should add suitable relationship names.

14.11 PG TOPIC: Different approaches to normalization

An interesting difference exists in the way some researchers describe normalization. The discussion in this chapter uses the four-column method found in most textbooks on SSADM RDA. The other way of normalizing we shall write about here is identical for the SNFs and TNFs. The works of Codd and Date, and of those who follow them closely, show a big difference in their approach to FNF.

These other writers do not initially form FNF tables as we have shown above.

---

[1]   HINT Linking a compound key entity with another entity can be said to have '*Small keys grabbing large keys*'. The way to treat foreign keys for the ERD can be optimized in the phrase: '*Crows feet grab the asterisk*'.

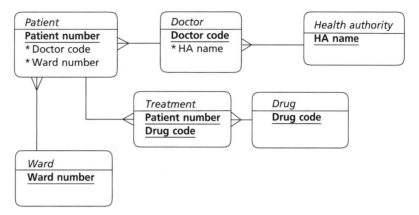

Figure 14.23 Entity relationship diagram generated for patient card and drug card.

Instead of simply creating the list of data items that we have called UNF, and transforming them into FNF, they deal with repeating groups in a rather different way. Their principle aim is to 'ensure that there is a single value at the intersection of each row and column' (Connolly and Begg, 1999, p. 201). Connolly and Begg (1999), for instance, describe this process, and they also show a procedure very like the one we have described, calling it the 'alternative first normal form'.

The advanced student could well research these two ways of producing FNF and evaluate which they think is the better method.

14.12 Tutorial 14.1

Using the data items in the Drug Card below perform normalization to TNF. The layout of your answer should conform to the SSADM standard. State any assumptions you need to make.

	Drug Card	
Patient number: 923	Surname: Face	Forename: Baby
Ward number: 10	Ward name: Greenfields	
Drugs prescribed:		

Date	Drug code	Drug name	Dosage	Length of treatment
20/5/88	CO2355P	Cortisone	2 pill × 3 day after meals	14 days
20/5/88	MO3416T	Morphine	Injection every 4 hours	5 days
25/5/88	MO3416T	Morphine	Injection every 8 hours	3 days
26/5/88	PE8694N	Penicillin	1 pill 3 × day	7 days

14.13 Tutorial 14.2

The report below indicates which customer will be serviced by which warehouse and where the warehouse is located. Using the report, perform normalization to TNF.

Handy Tool Ltd.

Monthly Salesperson Report

Salesperson number: 1001   **Date:** May 1999
Salesperson name: Bloggs, J.
Sales area: West

Customer number	Customer name	Warehouse code	Warehouse location	Sales
101	Smith & Co.	2	Bilston	12,950.00
102	Jones & Son	4	Oxley	21,255.00
103	Smithsons	7	Parklane	10,199.80
			Total monthly sales	44,4404.80

Proposed data requirements

- [ ] The data for the new system
- [ ] Rationalization and optimization
- [ ] The logical data module of the proposed system
- [ ] Tutorial 15.1

15.1 The data for the new system

We now have several tables in third normal form (TNF) as the basis of the database for the proposed information system (IS). However, there are a couple of things to do before we can be fairly sure that we have the best set of tables for the system. The tables should be *rationalized* and *optimized* before being made available to the newborn computer system. Furthermore, we can take the normalized tables and produce a new logical data model for them.

15.2 Rationalization and Optimization

We now deal with a topic that may seem odd. After going to all the trouble of normalizing the data for the new system, we shall eventually propose that it is often a good idea to de-normalize it a little.

15.2.1 Rationalizing the tables

The first step in this optimization of data in tables will not seem so unreasonable. After the relational data analysis (RDA) of the information on several different pieces of paper, more than one table will be created with much the same data items or attributes and certainly with exactly the same key. It is easy to think of a simple example for this. On one form, the details for a customer may have a customer number, the name and the address. While another form has customer number, name and credit limit. So, after normalization, two tables for customer details may be created: each will have **customer number** as key, but different attributes will be included under this key. A straightforward step to take is to combine all the attributes dependent on the key **customer number** in the same table.

This should be considered and carried out for all the tables produced by the RDA. There should be only one table for each key field, whether it be a simple or compound key.

There may also be hidden foreign keys. Foreign keys that are produced out of normalization should be obvious—during RDA we have, after all, spotted them and created another table for the dependant attributes. However, sometimes a table will contain an attribute that (almost accidentally) turns out to be the key field of another table. For instance, if a form contains the **customer number** as an entry, but it has no other fields associated with that customer number (such as customer name and address), the RDA will not have an asterisk against this customer number to show it as a foreign key. Yet, the odds are that somewhere we will have created a **customer** table with **customer number** as its key. Such hidden foreign keys should be sought out and pinpointed with an asterisk.

Another step in rationalization is to look at all the attributes in the tables and see if there are any data items with different names but have exactly the same meaning. Such attributes might be called *synonyms*, because they mean the same thing but happen to have different words to name them. Examples of such synonyms are legion. A *client* in one place is a *customer* somewhere else; and an *invoice* to some people is a *bill* to others. Potential synonyms should be checked with the user, and, if they really mean the same thing, one agreed term should be allocated to them.

15.2.2 Optimizing the performance of tables

There is a much less obvious reason for carrying out more work on the normalized tables, and, at times, actually to *de-normalize* them. One of the most evident results of RDA is that lots of tables tend to be created, and some of these tables can be very small. There is nothing intrinsically wrong with small tables, but they can lead to inefficiencies in processing.

Consider a table that contains only one record. Perhaps, in an online system, the processing requires that this table is updated for every transaction. To allow update access, a common practice of database systems is to lock the record so that other users cannot update it as well. This could mean that only one person at once is allowed to process any transactions in the system. One of us experienced such a system in practice. The table in question actually had two records in it, but one of those records was being updated for every transaction. It meant that, although several people were meant to be inputting data, only one could really do so. In such circumstances, it could well be better to hold the information on some larger table.

As another example, consider the course table in the example in Figure 14.21. It may be that there are only two courses in this system and that some data in this table needs to be updated for every transaction. It may be better for efficient processing to hold that particular data on every student record rather than the course record. So, the data would be *de-normalized* so that course information is held on every student record rather than having this separate course table.

Questions of processing efficiency could also lead to the idea that single tables ought to be split into two or three tables. So perhaps it would make sense, in some circumstances, to split the student table into students with surnames starting with letters between A and M, and another for those from N to Z. If this were done, then an IS that is being used by only two users could allow one to work on the lower half of the alphabet while the other works on the higher letters. So these users could, if necessary, lock the whole file they were working on without interfering with the work of the other.

15.3 The logical data module of the proposed system

Following the RDA of the data required in the new system, we can use logical data structures (LDSs) to illustrate its relationships. The resulting diagram will become the LDS or entity relationship diagram (ERD) for the **proposed** IS.

To begin with, we are armed with the set of tables that have come out of the RDA. All of these tables will have keys. Some of these keys will be simple, consisting of one attribute, while others will have complex keys: composite or compound. Still other tables will include, in their non-key attributes, foreign keys that point to other tables. It is the compound keys and the foreign keys that give clues about the way an LDS can be constructed from the normalized tables.

The structure of the compound key of **Module code** and **Student number** in worked example 3e, included in the previous Chapter as Figure 14.21 (reproduced here as Figure 15.1), shows that it is a link entity. The table is named **module registration** and links the two tables **module** and **student**. These three tables can therefore be shown in an ERD as shown in Figure 15.2.

One of the rules for modelling the normalized tables in an ERD can now be educed: the crow's feet in the ERD for a set of two entities and their link entity attach to the link entity.

We come now to dealing with tables containing foreign keys. In the worked example, there are several of these. The table **Module** contains the foreign key **Module leader code**, pointing to the **Module–Leader** table. There is also the attribute **Course code** in the **Student** table, which is a foreign key pointing to the **Course** table. These relationships can be shown in the expanded ERD shown in Figure 15.3.

A second rule shown here is that the *crow's feet grab the foreign key*, or *crow's feet grab the asterisk*.

Normalization – worked example (3e)

TNF – separate other attribute dependencies.

UNF	Level	FNF	SNF	TNF	Table names
Module code	1	**Module code**	**Module code**	**Module code**	
Module name	1	Module name	Module name	Module name	Module
Number of credits	1	Number of credits	Number of credits	Number of credits	
Module leader code	1	Module leader code	Module leader code	* Module leader code	
Module leader name	1	Module leader name	Module leader name	Number of students	
Number of students	1	Number of students	Number of students		
Student number	2			**Module leader code**	Module
Student name	2	**Module code**	**Module code**	Module leader name	leader
Course code	2	**Student number**	**Student number**		
Course title	2	Student name	Grade point	**Module code**	
Grade point	2	Course title	Result code	**Student number**	Module
Result code	2	Course code		Grade point	regis-
		Grade point	**Student number**	Result code	tration
		Result code	Student name		
			Course code	**Student number**	
			Course title	Student name	Student
				* Course code	
				Course code	Course
				Course title	

Figure 15.1 Final RDA for the module class list.

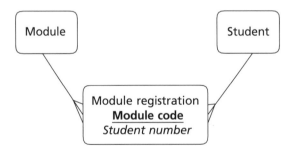

Figure 15.2 Entity relationship diagram part 1.

15.4 Tutorial 15.1

Create the LDSs for the other two worked examples in Chapter 14.

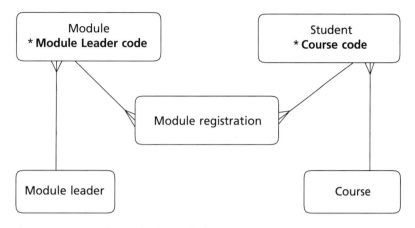

Figure 15.3   Final ERD for the worked example.

16 Processing requirements

- Processing in the required system
- Required logical and physical data flow diagrams
- Worked example for the video shop system
- Specifying the program
- The program specification
- Tutorial 16.1

16.1 Processing in the required system

The last chapter discussed the data for the proposed system, and later we shall deal with the design of the other outer part of the three-tier architecture for the required system: the user interfaces. In this chapter we consider the middle tier of processing. We need to discover what processes are wanted for the new system and we want to be able to specify them so that a computer programmer can code them. Finding out what processes are required necessitates the use of data flow diagrams (DFDs) for the new system.

The stages of DFDs so far explained have been:

- *the current physical DFDs*; and
- *the current logical DFDs*.

The current physical DFDs model the information system (IS) that is running at the moment—the old system; and the current logical DFDs show the underlying logical model of the same current system. At some stage we need to model the way the new system is going to work. There are, conceptually, two further steps in the full set of DFDs:

- *the required logical DFDs*; and
- *the required physical DFDs*.

So, we move from the current physical model to the required physical model, through two logical models: the current logical and the required logical. See the diagram for this idea.

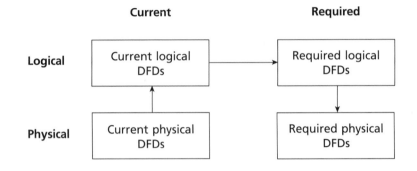

16.2 Required logical and physical data flow diagrams

16.2.1 Required logical data flow model

The logical DFDs for the required IS are derived from:

- the current logical DFDs;
- the requirements catalogue (RC);
- the selected business system option (BSO)

Diagrammatically, this can be shown as in Figure 16.1—note that *DFM* stands for the data flow model: being **all** the DFDs generated at the particular stage.

Harking back to Chapter 3, the RC is produced from the problems-requirements list (PRL), and, in turn, is incorporated into the various BSOs suggested for the system. One BSO is chosen for the IS containing, as it does, all its particular requirements from the RC.

The required logical DFDs are developed in a very similar way to the other sets of DFDs we have met. They have one Context diagram, one Level 1, and however many Level 2 and 3 DFDs that are deemed to be necessary. The designer needs to decide which processes are to be included in the required model. Generally, any process that could be carried out by a computer should be included as a candidate for automation, but any process that requires human action or judgement should not be included as part of the required logical DFDs.

Required system logical DFMs:

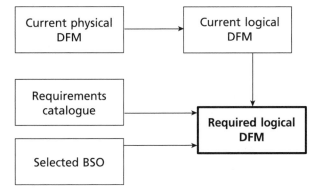

Figure 16.1 Deriving the required logical data flow model.

To develop these DFDs, look at the RC and decide what new processes, data flows and data stores are needed to meet the new requirements of the system. These components are then added to the current logical DFDs to produce the required logical DFDs. The data stores in the DFDs will be the ones agreed upon as a result of normalization, optimization, and rationalization.

16.2.2 Required physical data flow model

The final stage of the design of the processing of the new system is the production of the DFDs for the required IS. This final procedure does not seem to be carried out very often; most practitioners are happy to have constructed the required logical data flow model and use this for the processing of the new system. However, technically, this step is the closing stage of the development of the DFDs.

The required physical DFDs should look rather like the DFDs of the current system, but of course with some new processes, data stores, and flows. Some processes that are important to the new IS, but are not automatable, might be included again—after their exclusion in the logical data flow models. Also, the locations for the required processes could be added, and other physical attributes of the system such as paper forms, and (who knows?) filing cabinets, trays, and spikes.

Just as the current system DFDs can be compared with the current system entity relationship diagrams (ERDs), so the required data flow model should be compared with the required logical data model. The principles for doing so are exactly the same as the ones we described in Section 10.8. We repeat the summary of those principles here.

> ### Relationships Between DFDs and LDSs
>
> - Logical data structures (LDSs) will or do reflect the structure of stored data.
> - A DFD shows data moving about the system and being stored in data stores.
> - Each data store could represent one or more entities.
> - An entity may not appear in more than one data store.
> - A data flow consists of data items. Each and every data item should belong to an entity.

All these, except one, can refer to the comparison between the required LDS and DFDs.

16.3 Worked example for the video shop system

As an example, Figure 16.2 and 16.3 give the logical Context and Level 1 DFDs for a (more complete) current video shop IS.

Current logical context level DFD:

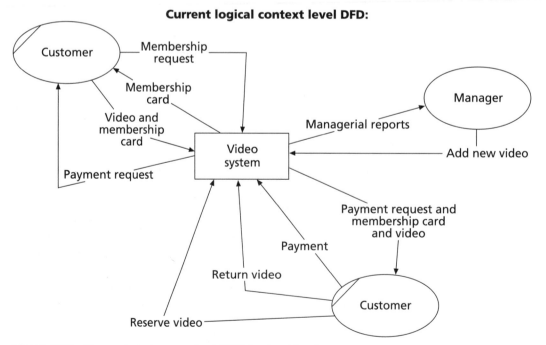

Figure 16.2 Current logical context level DFD for the video shop.

Current logical level 1 DFD:

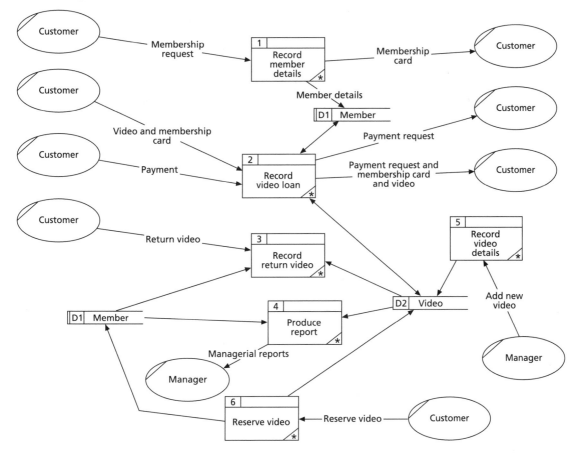

Figure 16.3   Level 1 current logical DFD for the video shop.

The relevant RC entries for the chosen BSO have these additional characteristics:

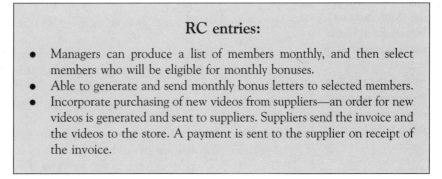

RC entries:

- Managers can produce a list of members monthly, and then select members who will be eligible for monthly bonuses.
- Able to generate and send monthly bonus letters to selected members.
- Incorporate purchasing of new videos from suppliers—an order for new videos is generated and sent to suppliers. Suppliers send the invoice and the videos to the store. A payment is sent to the supplier on receipt of the invoice.

Required logical context level DFD:

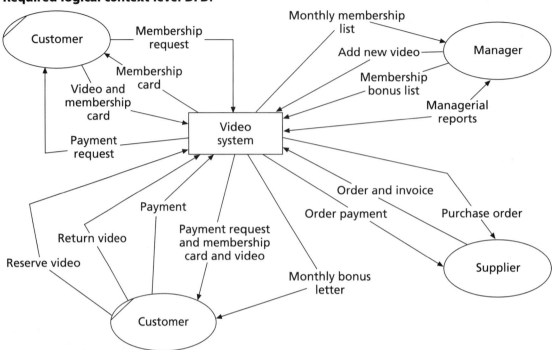

Figure 16.4   Required logical Context level DFD for the video shop.

Incorporating these needs into the data flow model might produce the Context and Level 1 required logical DFDs shown in Figure 16.4 and 16.5. Note the addition of the **Supplier** external entity and the **video order** store; also the inclusion of the bonus letters.

16.4 Specifying the program

Program specification   A precise statement of the effects that an individual program is required to achieve. It should clearly state what the program is to do without making any commitment as to how this is to be done.

DOC

Specifying the computer programs for the required system is a very important stage of the development process. It should not be the case that the completed required physical or logical DFDs, along with the data and interface designs, are handed over to a programmer or a group of programmers with no further guidance. Generally, each process in the DFD needs to be specified as a computer program. Much of the effort to do this is indeed in the collation of the relevant user interfaces and table designs; but we also need the formal program specification to tell the programmer what the program is intended to do.

Potentially, every lowest level process on the required system DFDs is a computer program. Computer programs can be broadly divided into two types: *enquiry programs* and *update programs*. In this rather simplistic view, enquiry programs are those that allow the user to obtain information from the system, while update programs are those that change data in the system. The required DFD can be studied to determine the category into which each process falls. Those processes that have a data flow to an external entity such as a customer

Required logical level 1 DFD:

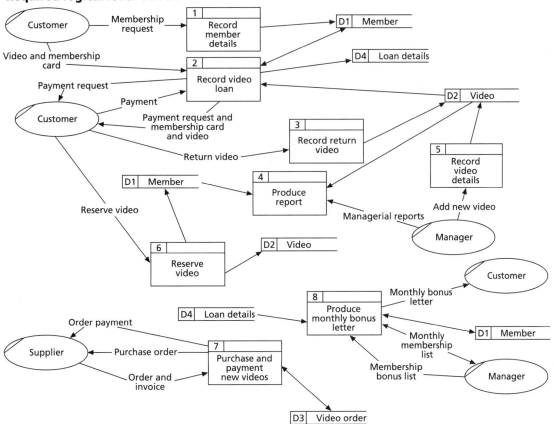

Figure 16.5 Required logical level 1 DFD for the video shop system.

or other user will be enquiry programs. Those that deal only with data stores are update programs. However, many processes output to users and interact data stores; these may also be classified as enquiry programs.

Whatever the classification of a program, it needs to be specified in order for a programmer to code it. The deliverable of this procedure is the program specification.

All programs can be seen as consisting of three parts: The inputs and outputs; the program processing; and the data it uses. Figure 16.6 shows part of the required logical DFD of a sales IS. Note that there are two low level processes that will require specifying. Process 1.1 is an update program and 1.2 is an enquiry program. In order to explain, in very general terms, what is to be expected of these simple programs, we shall look at each in turn.

16.4.1 Enquiry programs

Figure 16.7 shows a general diagram for enquiry programs.

Identifying update and enquiries

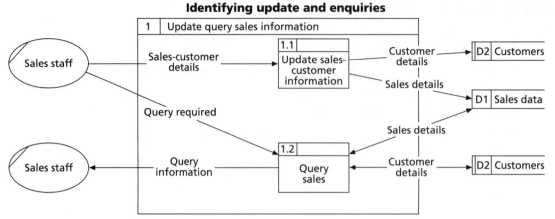

The DFD shows both an update (1.1) and an enquiry (1.2). Note the enquiry consists of two IO data flows.

Figure 16.6   Identifying processes for specification.

Enquiry program

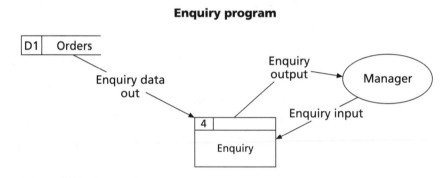

Figure 16.7   A general enquiry program.

In this general enquiry program, the external entity, **Manager**, puts in a request (often including the value of a key field). The program carries out some simple validation, such as checking the format of the input, and (if it is valid) looks up the record on the database. If the particular record is found, the program outputs the requested details on the screen for the Manager. If the record is not found, an error message may well be presented instead.

16.4.2 Online update program

Update programs come in a variety of shapes. Figure 16.8 shows a generalized online update program.

The updating of data often follows an enquiry on that database. Here, the user external entity inputs data to the process (update online). The bidirectional data flow locates the correct record on the database and updates it. After this, an output is made to the user to say that the update has been successful. Online update programs may add, delete, or amend the record in question. For each of these, the user inputs the required data, often including the key field details for

Online update program

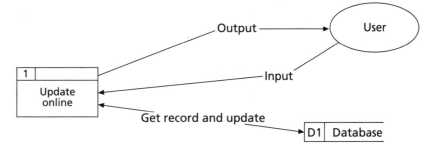

Figure 16.8 General online update program.

the required record, by altering data on the screen provided by the enquiry process.

For **deleting** a record, the program tries to find the record, but if it cannot locate it, it reports back with an error message—this means that the user is trying to delete a record that does not exist. If it finds the record, it deletes it and reports back that it has been successful.

When **amending** a record, the program attempts to locate it and, if it cannot, an error message is output to the user—as you cannot amend a record that is not there to change. Should the program find the record, it can be changed and the user is told that the attempt to alter the record has been successful.

For **adding** a record, the program still attempts to locate it. In this case, if the program finds the record there is an error—because the user is trying to add a record that already exists. If the record does not exist, then the details can be added to the database and the success reported back.

16.4.3 Data transfer program

A particular variant of the update program is the one that takes one or more files (tables) as input and transfers data from them to one or more output files. We call this a data transfer program. Figure 16.9 gives a particular example of this.

A data transfer program may be run online or batch and takes records from one or several files and outputs them to other files. A good example of this type of program is one that merges two files into one. While doing this it could be selecting records such as picking out all failed students from two or more course files and putting them onto one file.

16.4.4 Report program

A special sort of enquiry program is the report program. In this case, shown in Figure 16.10, records are taken from one or more input files and output (on a screen or on paper) for the user to see.

The report program could, again, be online or batch. The program reads one or more files, perhaps carries out some processing or selection, and outputs the report to the user. Figure 16.11 gives an example of a paper-based output report for the video shop.

Data transfer program

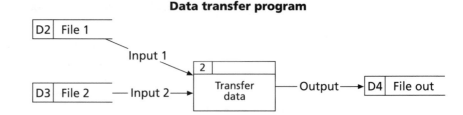

Figure 16.9   A general data transfer program.

Report program

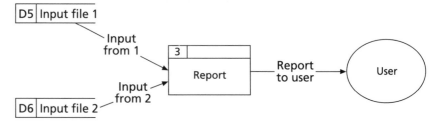

Figure 16.10   General report program.

Video Club Ltd					
Stock analysis by video type					
Week ending DD/MM/YY					

Video type	XXXXXXXXXXX				
Video number	**Description**		**Qty**	**Unit price**	**Stock value**
XXXXX	XXXXXXXXXXXXX		9999	£££.££	£££££.££
	↓	↓	↓	↓	↓
↓					↓
Total for types			999999	£££££.££	£££££.££
Grand total – All types			9999999	££££££.££	££££££.££
End of report					

Figure 16.11   An example report for the video shop.

16.5 The program specification

The specification for the program contains all the information that passes from the systems analyst or systems designer to the programmer. It therefore requires information about files, processing, and inputs and outputs. Often, the specification is set out in some standard form stipulated by the company. Figure 16.12 shows part of a specification for a program that amends details of the company's customers.

This example specification contains cross-references to the input–output (I–O) descriptions needed, the dialogue reference, references to the tables

Program specification entry		
System: Video club		
Type: update	Program name: Update 1	Program ID: vs33.1
Description: Amends customer details		
IO descriptions	1	Dialogue number:	DIAL 2.
Process number	4	Tables names	1. Customer

Logic and formula
While customers to alter
Do
Find record in Customer Table using Customer Number
If found
      Amend record
Else
      Set up error message
End If
Output message
End Do		

Comments:		

Figure 16.12 An example program specification.

Input–output description

Each file and data flow requires an Input–Output description.
Here is one from the program specification example:

IOD number	From	To	Data flow name	Data items	Comments
1	Customer	4	Customer details	Customer number Name Address Date joined Age	Key field

Figure 16.13 An example input–output description.

needed, and the processing logic set out in Structured English. Figure 16.13 gives an example layout for the I–O descriptions for the example specification. Dialogues, and other aspects of I–O design, are covered in Chapter 18.

The logic and formulae required for the program also needs to be expressed. This can be done in very similar ways to those we showed you for describing processes in Chapter 11: decision tables, decision trees, and Structured English. The aim of this part of the specification is to make the functions of the program explicit to the programmer. As we noted earlier in this chapter, it should say what the program is to do, and not precisely how to do it. We include now the logic descriptions for the types of program we have already discussed.

Enquiry: Structured English

```
Accept enquiry
Validate enquiry
IF record not there
   Produce error message
ELSE
   Access database
   Get answer
END IF
Output to user
```

The decision table version of the fundamentals of a general enquiry program is shown below.

Enquiry: decision table

C Record present?	Y	N
A Access database	X	—
Get answer	X	—
Error message	—	X
Output to user	X	X

The basic processing for adding a record to the database is shown below in Structured English. Note that the record is only added if the input is valid, the transaction is added, and the record (or rather the key value) is not present already.

Update: add

```
Accept update
Validate input
IF input valid
    IF add
        IF record not there
           Insert record
        ELSE
```

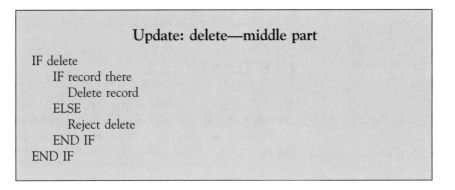

```
                Reject add
            END IF
        END IF
    ELSE
    Error 'Invalid input'
    END IF
    Output to user
```

For deleting a record, the Structured English below shows the rudimentary processing that deletes the record, but only if the transaction is delete the record is actually present to be deleted.

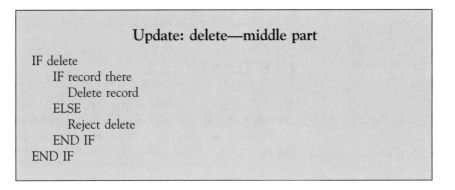

Update: delete—middle part

```
IF delete
    IF record there
        Delete record
    ELSE
        Reject delete
    END IF
END IF
```

Amending or changing a record can only occur if the record is already present, and the transaction is amend. In the Structured English below, the record is replaced with the altered information if the transaction is amend and the record is already there.

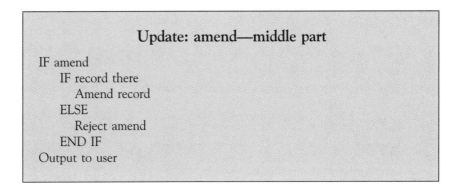

Update: amend—middle part

```
IF amend
    IF record there
        Amend record
    ELSE
        Reject amend
    END IF
Output to user
```

We show below a full decision table version of updating a database and amending, deleting or adding a record.

Update: decision table version

C	Valid input?	Y	Y	Y	Y	Y	Y	Y	N
	Add?	Y	N	N	Y	N	N	N	—
	Delete?	N	Y	N	N	Y	N	N	—
	Amend?	N	N	Y	N	N	Y	N	—
	Record there?	N	Y	Y	Y	N	N	—	—
A	Insert	X	—	—	—	—	—	—	—
	Delete	—	X	—	—	—	—	—	—
	Amend	—	—	X	—	—	—	—	—
	Reject	—	—	—	X	X	X	X	—
	Invalid input	—	—	—	—	—	—	—	X
	Output	X	X	X	X	X	X	X	X

16.6 Tutorial 16.1

Draw up a program specification entry, like the one in Figure 16.12, for the programs that add and delete records for the customer details table.

Draw the decision tree for the update process shown in the decision table above.

Storage of data

- Data storage
- Conventional files
- Types of files
- Databases
- Data codes
- The sizing of files
- PG TOPIC: sizing is more complicated
- Tutorial 17.1

17.1 Data storage

Data is at the heart of any computer system, or any other information system (IS). With any computer-based system, data has to be available when the user wants to use it. It must also be accurate and consistent.

A good IS will allow for efficient storage of data, enabling information from the stored data to be in a useful form for management, planning and controlling purposes, as well as for decision making. Other objectives also include the efficient updating and retrieval of data.

We shall consider two main approaches to the storage of data, called *file organization*: conventional files and databases.

17.2 Conventional files

There are a great many computer systems out there in the world that are utilizing conventional files. Many are *flat files* or *serial files*, where the information has little structure and is stored in the order it happens to be collected. These sorts of files include the word processing file into which this chapter is being keyed as well as the raw data that represents, for instance, the data for the hours worked by a group of employees. More structured files are those called *sequential files*, *random files*, and *indexed sequential files*.

17.2.1 Sequential files

> **Sequential file organization** The rows in the file are stored in sequence according to a primary key valve.
>
> *Huffer et al., 1999*

These are files in which the records are organized into the order of their key field or fields. They are sorted in the order of, or in the *sequence* of, that key. These files may be stored on magnetic disc or on magnetic tape. If the storage medium is magnetic tape, then the only way a particular record can be accessed is by looking at all the records in order until the computer comes across the record in question. This *sequential processing* has been the traditional method of processing data, and, as a consequence, it is still commonly seen in older systems, but it has many uses too in modern IS. It is useful where the processing needs a large proportion of the records on the file to be accessed. This is referred to as processing requiring a *high hit rate*. ISs are replete with examples of processing that will access all or most of the records on a file.

When gas bills are processed, pretty well all the customers' records will need to be accessed in the order of their customer number. There are also many occasions when a bank needs to access all its customers in the order of their account number, for instance, when monthly statements are being printed out before posting.

17.2.2 Random files

> **Hashed file organization** The address for each row is determined using an algorithm.
>
> *Huffer et al. 1999*

You may also see these referred to, particularly in US publications, as *hashed files*, or *hashed random files*. In this file organization, which has to be on magnetic disc, the way the records are stored physically appears to be all over the place, and is therefore called *random*—but it isn't really. The position of a record on the disc is determined by a mathematical formula or *algorithm*—the so-called *hashing algorithm*. This algorithm is constructed so that records are, as far as possible, spread evenly over the whole surface of the disc. A computer program places a record in its correct position by running the key field of the record through the algorithm and positioning the record in the place dictated by the solution to that calculation. When the record needs to be accessed again, a program discovers the position it is to be found in by using the same mathematical formula.

This way of accessing data is called *random access* and is a very quick way of finding a particular record. The record is accessed not sequentially but *directly*, so this method is also called (perhaps more clearly) *direct access*. In sequential access, all the records before the record needed may have to be read before finding the one record that is wanted. In direct access, as long as the value of its key field is known, a record can be found immediately. However, for accessing that requires many of the file's records to be found in key order, a random file will prove to be very slow. So, random files are useful for processing that needs a *low hit rate* such as seat booking and reservation systems.

17.2.3 Indexed sequential files

Index   A table used
to determine the
location of rows in a
file.

Huffer et al., 1999

This very popular type of disc file organization is perhaps best approached by way of a short discussion about *indexes*. An index is a special part of a file that contains the values of the key fields of all the records in the file and an indication of where those records are to be found on the disc. This is very like the index of this book, which contains a list of words and phrases in alphabetical order alongside the number of the pages in which that word or phrase is to be found. In a disc index, the values of the key fields of all the records are held in key order, together with the place on the disc in which the records are to be found. The software accessing the records must look up the key field in the index and find where on the disc the record can be found. Similarly, whenever a record is placed on the disc, its position must be entered into the index along with the value of its key field.

A file that has an index can be called an *indexed file*, but this says nothing about the way the data records of the file are stored. Theoretically, but not often used in practice, the data records could be in any order or the order in which they were collected. The index will still be able to tell the software where each record is to be found. More commonly, an index can be set up for a file that is already in one sequence, so that the records can be accessed in another sequence. An example of this might be for a customer file that is held in surname order. That order may be fine for much of the processing required of it, but at times it will need to be accessed in customer number order. The same records can be kept in their same places on the disc, but, with a new index in customer number order, the file can be accessed in this different way. The index can also be used to access each record through direct access, similar to the random files mentioned above. Direct access is accomplished by searching the index for the value of the key field of the required record and going directly to the position on the disc indicated by the index. In computer systems such as the IBM S38 and AS400, all files are kept in the order in which the records are collected and all files have an index or several indexes so that they can be accessed sequentially or directly.

The term *indexed sequential file* (or *index-sequential file*) is properly reserved for files that are organized with the use of an index, but where the data records of that file are in sequential order like the sequential file organization. The order of the key field values is the same as the order of the data records so that the records can be accessed both sequentially and directly. This has been shown to be a very useful way of organizing files because it gives the advantages of both sequential files and random files. The data records can be read in their sequential order for processing that has high hit rates, and they can also be accessed directly for low hit rate processing. Examples of this are very common and include those systems in which only a small percentage of the records are updated throughout the day (low hit rate), but where the whole file needs to be printed in order (high hit rate), say in the evening or at the weekend.

Additionally, and also often found to be of great usefulness, the file can be accessed sometimes directly and sometimes sequentially by the same program. For example, with a file organized in alphabetical order, a processing requirement may be that only those records within the range G to L need to be printed. Rather than reading all the records for A to F first, the index can be used to discover where the first record with G can be found and accessed directly. Then all the records from G to L can be read sequentially in alphabetical order, because that is the way they are put on the disc.

To pay for what may seem to be the file organization that offers the best of both worlds, do not forget that the index takes up space on the disc. Sometimes, when records contain very little information except for the key field, the index can be almost as large as the space required for the data records themselves. This means that the file would be almost twice as large as might have been thought necessary. Also, the direct access of records is not as efficient as that achieved by a random file. A computer calculation using a mathematical formula is very quick, while looking through a disc index is, by comparison, extremely slow.

17.3 Types of files

Files come in a variety of types for a variety of uses. Perhaps the most important type of file is the *master file*. A master file is the file that contains the permanently held data for the system (or part of a system) and it is the data on this file that is updated in order to keep a record of the latest state of the data. Another type of file is the *transaction file*; this contains the data that is to be used to update the master file. Typically, the master file contains information such as the total hours worked by a company's employees (as well as a lot of other information), while the transaction file may contain the hours worked by the employees today or this week. The transaction file data is used to bring the master file data up to date. Other types of files are *reference files*, which are, in effect, lists of useful information to be accessed by a program during processing; and *archive files*, which are kept for long periods of time but are accessed rarely—they often show the state of a master file during different periods of its life such as last month or last year.

17.4 Databases

This is not a book about databases or database theory, but some detail of this topic is needed even for a book on basic systems design. There are many books on databases that the reader is urged to consult for a deeper discourse. We shall consider only those file organizations called *relational databases*. There are other types of databases, but we shall not be discussing them here.

A database has often been defined simply as 'a collection of data'. Following this definition, all files (including all the conventional files we have spoken about already) could be called databases. However, the term is generally used to refer to data that is accessed through a *database management system (DBMS)*. The *DOC* definition covers both these uses.

17.4.1 A few facts

A database is a centralized data store and not simply a collection of data or files. The data is intended to be shared by many users and many applications. The aim

---

Database

1. Normally and strictly, a body of information held within a computer system using the facilities of a database management system.

2. Occasionally and colloquially, a collection of data on some subject however defined, accessed, and stored within a computer system.

DOC

is to enable data to be entered, stored and updated once in one place. The physical data storage should be invisible (or *transparent*) to the user.

A *relational database* stores data in tabular form where each file represents a table and every field (attribute) is a column and each record is a row. These tables are related by common attributes such as composite key fields and foreign keys.

17.4.2 Advantages of a relational database

This type of database has the potential to evolve and is often more flexible than conventional files. This allows for it to change with the needs of the users. It should provide accurate and consistent maintenance of the data and provide for the efficient handling of *ad hoc* queries. One of the other objectives of the database is to ensure that all data required for current and future applications will be readily available. Data redundancy is eliminated (or reduced as far as possible) by designing the database using relational data analysis (RDA). As we have seen, this takes the views of the user (often ascertained by studying the data in the current physical system) and transforms them into less complex structures and reduces the duplication of data.

17.5 Data codes

We now move from the organization of files to the way data can be included in the records of those files. Data space has always been a problem for computer systems, so anything that reduces the volume of the space required for the data is to be encouraged. The coding of data can often help with this aspiration; it also cuts down on the amount of data to be input, and it helps to facilitate the validation of input data so improving accuracy, while increasing the control of the data.

17.5.1 Coding design

A data code is an ordered collection of symbols designed to provide unique identification of an entity or attribute. It can help to keep track of data items, classify information, conceal information, reveal information, and even to record and request an appropriate action.

Whatever type of code system is used, the codes must be concise, stable, unique, sortable, understandable, meaningful, and adaptable. *Conciseness* is important because one of the aims of coding is to reduce the data held and input. *Stability* refers to the fact that the code should be able to cope with all the data it needs to, without failing. A code system should produce codes that are *unique* so as not to refer to two different things. Connected to uniqueness is the idea that most codes need to be constructed so that they can serve as keys and can be *sorted* on those keys. Codes that are *understandable* will not be so complicated that people will get them wrong, as will those that are *meaningful*. Finally, an *adaptable* code system will be able to cope straightforwardly with small changes to the data such as the addition of new items.

17.5.2 Tracking items

The use of sequential codes can be used, for instance, to keep track of invoices. A sequential invoice number (101, 102, 103, etc.) may help in ensuring that an invoice has not been lost. As discussed earlier in this book, a customer account number is a better way of designating a customer than names and addresses. Such an account number may be constructed from other information held about the customer. For example, the name J Smith, who has the telephone number 516233 and a post code of TS11 4QQ could be used to construct a unique customer account number such as TS41-SMIj-516.

17.5.3 Classification of information

Data codes are often used to provide a meaningful classification system for the data. *Block sequence codes* split up a sequential code onto several blocks, each of which can refer to something important to the business. For instance, stock items can be allocated stock numbers in blocks to give a good idea about what sort of stock is being referred to. In a coding system using four-digit stock codes, household items can be allocated codes from 1,000 to 2,999, electrical items from 3,000 to 4,999, and clothes from 5,000 to 6,999.

The *Dewey decimal system* for library books is a version of a block sequence code that classifies books by their subject. Given the particular code of a book, a librarian (and a knowledgeable reader) can tell what topic the book is about. Also, readers can look for books on a similar subject by looking through the books in the catalogue with similar Dewey decimal numbers.

17.5.4 Concealing information

Codes can also be used to hide information for reasons of security. Ordinary text (*clear text*) can be encoded (or *encrypted*) so that those who are not authorized to read it will be unable to do so. Also, pricing information can be encoded to keep it unclear to prying eyes. The reader should consult the many textbooks on data encryption if interested in this subject.

17.5.5 Revealing information

Faceted codes are made up of meaningful parts. In a clothing business, the code for a stock item may be made up of three parts; for instance, the garment colour, its type, and its style. A three-character code could be made up for the clothes so that the colour is P for pink, R for red, G for green; type is D for dress, S for skirt, and H for hat; while style could be A for adult, B for baby, and C for child. A particular item of clothing could then be given the code PDB for a pink dress for a baby, and GHA is a green hat for an adult.

Mnemonic codes are often used to reduce the data held and entered while giving clues to the meaning of that data.

Hospitals in the UK have a three-character code that is close to the full hospital name, so that WNX stands for Wolverhampton New Cross hospital.

> **Mnemonic** is taken from the Greek *mneme* for memory, *mnēmōn* mindful, and *Mnemosyne* the Greek Goddess of memory and the mother of the muses.

Railways stations also have a three-character mnemonic code, so that BNS is Birmingham New Street station.

17.5.6 Codes to record or request actions

Often a code is used to represent the action that has been taken or should be taken about some aspect of the business. For example a simple one-digit code can be used to show the state of items on an order: 0 can show that the order item needs to be processed, 1 can represent that an item has been dispatched, 2 that it has been delivered, 3 that it has been returned, and 4 that it has been stolen. So, on the record for a particular order (or order-line), presumably keyed on the order number, each item ordered would have the one-digit code to show what has happened to it, or what needs to happen.

17.6 The sizing of files

A very important task that needs to be undertaken in the design of a new computerized IS is calculating how much space will be taken up by the data in the system. Simple file sizing requires only simple arithmetic. Its basis is to work out how many bytes are needed by each field in a record, and then to determine how many records are likely to be held in that particular file or table. Multiplying the bytes by the number of records gives the size of the table. Adding the sizes of all the tables in the system gives the size of the whole database. A supplementary consideration must be the addition of a contingency to give slack capacity to account for unanticipated factors. Often, 10 or 15 per cent is added to the calculated size of each table.

17.6.1 Worked example 1—a customer table

In the video shop, for a maximum of 100 customers, the size of the table holding the details of those customers can be calculated as follows:

Field name	Format	Length in bytes
Customer number	4 digits	4
Name	30 characters	30
Address	3 digits	
	22 characters	25
Date joined	dd/mm/yyyy	10
Date of birth	dd/mm/yyyy	10
Total for record		79 bytes
For 100 records		7,900 bytes
Plus 10% contingency		790 bytes
Total for table		8,690 bytes

17.6.2 Worked example 2—a video table

Also, for the table that holds details of the videos:

Field name	Format	Length in bytes
Video number	4 digits	4
Video table	20 characters	20
Category	1 character	1
Date bought	dd/mm/yyyy	10
Classification	2 characters	2
Total for record		37 bytes
For 100 records		3,700 bytes
Plus 10% contingency		370 bytes
Total for table		4,070 bytes

These are in fact very small files. Consider those that have records several hundreds of bytes long, for customer bases of several thousands, or hundreds of thousands. Also think about files that hold data for several years. Such files are very large and can become very much larger as time goes on. A long-term student file with 10,000 records of 1000 bytes each is 10 million bytes long! This is the sort of file size that industry and commerce ISs handle every day.

17.7 PG TOPIC: sizing is more complicated

This simple calculation of file sizes is fine as a beginning, but life, as is so often the case, is rather more complicated than that. A major complicating detail is the fact that files very often consist of more than just their data. A glance back at the start of this chapter will remind you that many files, *indexed* and *indexed sequential files*, have data **and** indexes. The calculations above are useful for dealing with the data in the files, but, if the file organization is not conventional but indexed, the index must also be taken into account.

The index for a file typically consists of the key values from the data records of the file set out in the order of the sequence of those keys. If all the keys for all the records are stored, as they must be for certain computers, then a considerable extra overhead needs to be catered for. Consider a file of 10,000 records each with a 20 character key field (or combination of fields). The index size would need to be 200,000 characters, or bytes, in addition to the size of the data records alone. Some files, such as link entity files, consist of little more than their key fields, so if such a file is held as an indexed file, then the space needed for index will be almost as large as the space needed for the data. The file will be **twice** the size of the data on its own.

To moderate this extra size for the indexed file, indexes may not have an entry for every single data record. There may only be an entry for the highest key field in a whole block of records. This cuts down on the size of the index considerably.

Random files and indexed sequential files have another complication. Often, these file organizations are chosen because the files are expected to grow a great deal. Space for this growth, which may also be a factor of other file organizations, must therefore be built into the calculations for the size of the file. This is not the paltry 10 or 15 per cent for contingency factors, but perhaps 100, 200, or 500 per cent for the expected growth of the file.

17.8 Tutorial 17.1

The video loan table for the video shop IS needs the fields in the following table.

Field name	Format	Length in bytes
Video number	4 digits	
Customer number	2 characters	
Date out	dd/mm/yyyy	
Date due	dd/mm/yyyy	
Payment	££.pp	

Assuming that 30 videos are taken out every day and that the shop is open 360 days in the year, work out the size required for the table, and include a 10 per cent contingency.

How big would the file be if it had an index consisting of an entry for every data record of the compound key made up of **video number** and **customer number**?

How big would the file need to be if it were expected to grow by 300 per cent during its life?

18 User interfaces

- ☐ Designing the user interface
- ☐ Methods of interface design
- ☐ The elements of interface design
- ☐ Considerations for good screen design

18.1 Designing the user interface

User Interface The means of communication between a human user and a computer system, referring in particular to the use of the input/output devices with supporting software.

DOC

Generally, the person making use of the new computer information system (IS) (the *user*) only ever sees that part of the system we call the *interface* or *user interface* (UI).

The UI is the way the user interacts with the system because it is the point at which data is input and information is output. The devices used for this (the *media*) include visual display units (VDUs), public information displays, and good old paper. For this discussion we shall be considering *menus*, *dialogues*, and *reports*.

Because, in the eyes of the user, the interface can define the success or failure of an IS, it should be as carefully designed as any other part.

18.2 Methods of interface design

There are several ways in which the interface can be designed. We discuss three here:

18.2.1 Paper based

The traditional method of interface design uses print layout charts, screen format sheets, and diagrams of menu hierarchies. The print layout chart is generally a piece of paper divided into many small boxes. This grid needs to be as wide as the paper on which a computer report is to be printed: perhaps 120, 132, or even 160 columns. It also needs to be deep enough to contain the number of lines on the printed page: perhaps 60 or 80 rows. The point of the grid of boxes is that it allows the designer to put every character that is to appear on the printout in precisely the correct place.

The screen format sheet is similar but smaller. This is often only 80 characters or boxes wide and 20 or so deep. It represents the VDU screen and allows the designer to set out each character on that screen in the place it belongs.

We shall discuss menu hierarchies later.

18.2.2 Menu, dialogues and report developer tools

There are software products available that allow the development of menus, dialogues and reports (both screen and on paper) to be developed online. Often this software will also generate the program code necessary to produce these.

18.2.3 Database and 4GLs

Many database management systems contain report generators to produce simple reports with very little effort. Also, many *fourth generation languages* (4GLs) have facilities for simple screen and report production.

18.2.4 Prototyping

UIs are most successfully developed using a prototyping approach. An initial (first cut) interface is generated by the designer and shown to the users. The users criticize this and a better version is produced. This process continues until the users are satisfied with the way the screen and report looks. Often, many such iterations are necessary before the interface design is agreed.

Pitfall

Users must be made aware that the interface they have been involved with is only the surface of the computer system. There is a big danger that once a user has become satisfied with the interface, they think that the entire system has been completed and want to use it straight away.

18.3 The elements of interface design

The three elements of the design of the user interface discussed here are:

- *Menu design*: the design of how the user will navigate through the system.
- *Dialogue or screen design*: the design of the screens and/or windows that allow the user to view and edit information.
- *Report design*: the design of the printed or displayed reports produced by the system.

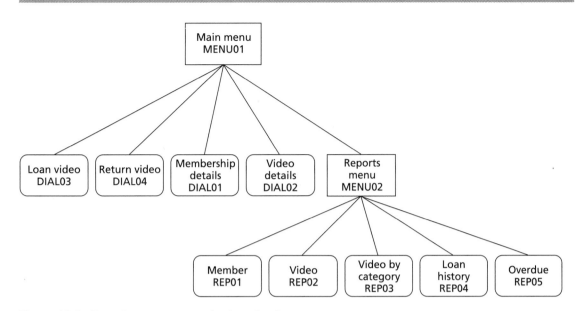

Figure 18.1 Example menu structure for the video shop.

18.3.1 Menu design

Menu design defines the way in which a user will *navigate* through the system. Users need to know how to get where they want to be from where they are now—navigation refers to the routes by which they can do so. Different users with different roles, such as keyboard operators, managers, and customers, may well require different menu designs. Each may, for example, need to be able to access different parts of the system. The customer will probably not have the same access as the manager.

A common way of designing menus is by a *hierarchical menu design*. This can be represented using *menu structures*. A menu structure is a diagram that shows how the interface is arranged. For the video shop computer system, the menu structure may well look like Figure 18.1.

You should be able to see that the structure is set out not unlike structured programming diagrams or entity life histories (ELHs). The top of the structure is the main menu; this is what is seen when the system is switched on or entered. On this screen a menu is presented with five different options, some of which lead to *dialogues*, and others that lead to other menus.

The list of options that we call a menu is not dissimilar to a menu in certain restaurants. It is a list of actions that we could ask the IS to take, and from which we choose one. In the restaurant, the action requested is for them to cook a particular meal. In the video system, the five items to be selected from here are: loan video, return video, membership details, video details, and reports menu. The first four of these are dialogues and the last leads to another level of menu. A dialogue in an IS is the conversation that takes place between the computer system and its user. Typically, the user is asked a question and is allowed to

Menu A list of options that may be displayed either vertically or horizontally on a screen and from which one or more items may be selected using an input device.

                                    DOC

Dialogue The sequence of interaction between a user and a system.

                    Hoffer et al., 1999

input and answer. This may not seem like much of a conversation, but that is the reason it is called a dialogue.

Note that, in Figure 18.1, dialogues are put in rounded boxes and menus in square boxes. Each has a unique reference code as well as a meaningful name. So the Main Menu is MEN01, and the Loan Video dialogue is DIAL03.

Traditionally, the way the user navigates around a hierarchical menu structure is by starting at the top and descending to the next level by making a selection. The way to get to another dialogue or menu is to return to the previous level and select again. So, in the example, once the user has finished inputting some loan details in DIAL03, they must return to the main menu in order to, say, select the dialogue for inputting video details (DIAL02).

On the other hand, the system could be built so as to allow users to go from the dialogue they are now in to the next dialogue they require, without going to the bother of returning to the previous level in the hierarchy or perhaps back to the top most menu. So, by selecting a button in the DIAL03 dialogue, the user could go straight to the DIAL02 dialogue without seeing the MEN01 menu again. This form of navigation can be called *functional navigation*, because the user goes directly to the function they require from wherever they are in the system. Another example would be the functional navigation from Membership Details DIAL01 to Loan Video DIAL03. After a new member's details have been input in the Membership Details dialogue (DIAL01), instead of returning to the main menu, the user can select a button that takes them directly to Loan Video (DIAL03).

Functional navigation could be the only method of getting around the system, or it could be combined with hierarchical menus. Our suggestion would be that hierarchical menus are a good way to design the system and a good way for new users to become familiar with it. However, hierarchical menu systems with many levels can become irritating to the experienced user and very annoying to the expert. So systems should also incorporate some functional navigation to allow non-novice users to dive about the system without negotiating the full menu structure.

Hint

- Hierarchical menu structures for novices.
- Functional navigation for experienced users and experts.

18.3.2 User roles

As mentioned above, different users need different things from the same system. The menu system can be used to provide different *views* of the IS. The view of the system describes the way a particular user sees the system. A customer sees only those aspects that concern them, while the system user may be allowed to access other facilities and the manager might have access to the whole thing. If different views are **not** provided, so that everyone sees everything, there may be serious security implications. Our advice is that it is better to tailor the menu to each user role. For example, the video store manager will require access to all dialogues and screens whereas an operator only requires access to issuing a video.

So a menu structure should be developed for each role, and each of these structures can be entered by a user identifier and password when they log on to the system. To make things a little easier for the developers, common dialogues can be used by more than one user, and menu items may well appear in more than one user's menu. For example, the video shop manager and the sales staff may

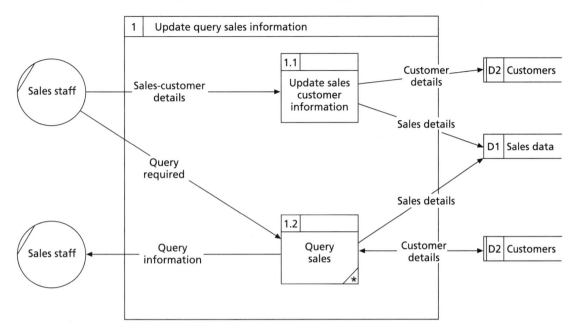

Figure 18.2   A DFD to illustrate the need for user interfaces.

both need access to the Loan Video dialogue (DIAL03); so this entry could appear in both their menu structures.

18.3.3 Dialogue design

*Graphical User
Interface* An
interface between a
user and a computer
system that makes use
of input devices other
than the keyboard and
presentation
techniques other than
alphabetic characters.
... Perhaps the best-
known GUIs are those
used on Microsoft
©Windows PCs and
Apple Macintosh
computers.

DOC

Dialogues are the way users interact with the IS, generally through a VDU screen or window. The dialogue is used to input data and output information. It may be text-based or use a windows-type *graphical user interface* (GUI).

The dialogues that are needed are identified from the required system dataflow diagram (DFD). User interfaces are possibly required between all external entities and the system, and especially likely are those between human external entities and the system. Furthermore, a dialogue may well be needed for the process that receives data from or sends data to the human external entity. As an example, Figure 18.2 shows a Level 2 required logical DFD. This system could have a dialogue for each input–output data flow that interacts with a human external entity. In this case, one dialogue is required for the Sales Customer Details data input as well as one for the Sales Query covering both the Query Required input and the Query Information output flows.

18.4 Considerations for good screen design

We include several factors that ought to be thought about in order to produce high-quality screens.

18.4.1 Layout

The screen should not be cluttered with too much squeezed on to a screen or window. If the screen is to be used to input data from a paper form, the screen should be set out in the same way as the form. Each screen or window should follow a consistent layout and any company conventions or standards. Also, if different parts of the system work in different ways they will confuse the user and lead to errors in the way the system is used.

18.4.2 Navigation

Pitfall

Do not force the user to fill up a field that does not need to be filled (e.g. an address field) in order to pass to the next. Consider carefully whether the user should be made to move from field to field by using a mouse. This can become very irritating and slow the work considerably.

Careful attention should be paid to the order of the fields on the screen and the way the user will navigate between the fields. The user may move from field to field by using a mouse, a tab or return key, or automatically when the field is full.

18.4.3 Use of control buttons and function keys

Think carefully about the location and effect of control buttons and function keys. Commonly, such buttons include a button to return to the previous level of the menu hierarchy, a help button, and a submit button. Function keys can be used to search, scroll, escape the system, and such like.

18.4.4 Headings

The top of the screen should have the system identification, and dialogue identification.

18.4.5 Data fields

The fields that are to contain the data input and the information output must be placed carefully. Consideration must be given to the type of field, its justification (left or right), and its length. The type of field is generally *alphabetic*, *numeric*, or *alphanumeric*. Alphabetic fields contain only alphabetic characters and spaces; numeric fields hold numbers and decimal points; and alphanumeric fields may hold any printable character. A field may need to hold a currency amount or a date.

Each input field will probably require a label; this is some text that identifies the content of the field. Common field labels are **Name**, **Address**, **Date of Birth**, and many others.

18.4.6 Validation

Often, the IS will be required to validate the data that is being entered. Dates can be validated to conform to date format, and may have to be before or after today. Many other fields could be checked for correctness. Numeric fields may need to be within a particular range, and even alphabetic fields may need to be, say, **Y** or **N**.

Pitfall

Students (and too many practitioners) assume that fields like name and address are alphabetic. In fact many names have characters other than alpha characters and spaces. The **O'Tools**, **St. Johns**, and **Cholmondely-Warners** have apostrophes, full stops, and hyphens—all non-alphabetic characters. Also, many non-English names have accents above or below letters.

Place names in addresses, even in Britain, may contain full stops, hyphens and apostrophes too. Examples are **Ashby-de-la-Zouche, St. Peter's Road**, and **Westward Ho!**

Once management realize that it's not really 'the computer's fault', but the designer's, developing a system that forces names to be held as something other than their reality may lead to the exit door from your company.

18.4.7 Error and help messages

The user will need to be informed when a piece of data has been entered incorrectly. So error messages will need to be built into the screen design. These may appear in a standard place on all screens (say at the bottom), or they might appear next to the field in error. ©MS Windows allows context sensitive help as you move the mouse pointer over the item concerned.

18.4.8 Font, colour, and style

Different fonts, colours and styles can be used to highlight information. The colour and style of the background need to suit the application and not detract from the user's task. Note that too many colours can be ugly and confusing. We are aware of at least one computer system that almost failed because the input form was so garish that the keyboard operators could not stand to look at it.

18.4.9 GUI input

© Windows provides features that are not available on simple text-based screens. These include selection boxes, buttons, scroll bars, sub forms, and radio buttons. Icons may be used in place of text, but if their meaning is not obvious they can lead to confusion.

19 System installation

- Implementing the design
- Installation
- Planning the installation
- Changeover methods
- Tutorial 19.1
- Tutorial 19.2

19.1 Implementing the design

The term *implementation* is a sneaky one, having at least two meanings. The one we are using here is in the box.

At this stage of the game, the information system (IS) has been analysed and designed, and so have the files, interfaces and programs. But this is, so far, all design—nothing that actually works has been produced. The implementation of the system design is the generation of the working system and putting that system into use. For a computerized IS, the most obvious working elements of the system are the computer programs. This book is not concerned with programming beyond the discussion of program specification described earlier. The program development stages cover the detailed design of computer programs, their coding, and their testing. Like *Pooh Bah* in 'The Mikado'[22], this book is concerned with just about everything else, but not that.

> **Implementation**
> The activity of processing from a given design of a system to a working version ... of that system.
>
> *DOC*

19.2 Installation

However, also included in the implementation phase is the way the new computerized IS is to be put into use—or *put live*. This is the *installation* of the new system.

> **Installation** The organizational process of changing over from the current information system to a new one.
>
> *Hoffer et al., 1999*

---

[22]   Pooh Bah is the *Lord High Everything Else* in the Gilbert and Sullivan musical.

Once again, this term can be used to mean other things, but this is the definition we are using here. As suggested in the definition, another term for this is the *changeover* from the old system to the new one. At some stage, the company must be told to 'screw your courage to the sticking-place'[23], discard the old system, and transfer to the new one. This is a very scary time for any company, particularly if its continuation depends upon its computer systems. We shall also be considering under this heading the installation of the hardware for the system and the writing of the system documentation.

19.3 Planning the installation

> **Installation plan** A schedule that describes all the tasks that need to be carried out in order for changeover to be completed successfully.

Given the frightening nature of putting the new system live, you would expect us to advocate its very careful planning—we shall not disappoint. A well-thought out installation plan needs to be drawn up.

The installation plan is part of the overall project plan, and will be shown in the total Gantt chart for the project. Many activities have to be carried out in the installation. All the infrastructure must be in place such as the cabling, the hardware for the system, and, of course, the database and the software. This planning has to be done a long way ahead because of all the things that could and will go wrong. There will be possible delays in purchasing items and having them delivered. The cables for the new system will take a long time to be put in, and so will the rest of the hardware. Installing the software and database will also not be a simple process. Those who are thinking, 'How hard can it be?' to install software, are simply not aware of the way life conspires against you. Like the boy on the burning deck, they are just not aware of how bad it can be. (And, by the way, he died!)

19.4 Changeover methods

There are four major ways of moving from the old IS to the new one: *direct* (or *big bang*), *parallel*, *phased* (or *staged*), and *pilot*. We shall consider these in order, describing each and discussing their benefits and their drawbacks.

> **Direct installation** Changing over from the old information system to the new one by turning off the old system when the new one is turned on.
>
> *Hoffer et al., 1999*

19.4.1 Direct changeover

At a scary time, this is perhaps the scariest way of changing over to the new IS. At a pre-arranged time, the old system is switched off and the new system is turned on.

The changeover happens almost immediately (or, at least, overnight), so it is also called the *big bang changeover*.

---

[23] Shakespeare's Lady Macbeth.

Disadvantages

If the direct changeover is done exactly as the name suggests, there will be no going back afterwards. It is a high risk option with no fallback position and no safety net. Therefore, if serious errors are found that have an immediate effect on the company, this can be very dangerous indeed. For failure of a system that is critical to a company could destroy it completely. However, sometimes there seems to be no choice except to change over in this big bang way. Needless to say, the company must be very sure that the new IS will work pretty well perfectly before entering upon this action, and it will require meticulous planning.

Advantages

As will become apparent, the direct changeover is often the cheapest of the methods described here. Also, because of its terrifying nature, it can cause the developers and users to be very motivated, very committed, and very careful. Nothing concentrates the mind more than knowing you may be executed tomorrow. The timescale for the big bang is generally much shorter than for the other methods, so the benefits of the new system will be gained more quickly. Also, it provides a clean break from the old system with its problems and reputation. The timescale may however be increased because the fear associated with the direct changeover can give rise to very extensive testing of the system in order to ensure, as far as possible, that it will work properly. And, surely, this can be no bad thing.

19.4.2 Parallel changeover

> **Parallel installation**
> Running the old information system and the new one at the same time until management decides the old system can be turned off.
>
> *Hoffer et al., 1999*

In this method, the new system is run alongside the old one for the period of the changeover.

This period may be a few days or several weeks—even if things go very well. While the two systems are being run like this, the performance of the new one is compared with that of the old. Once the developers and management are happy that the new system is doing what it ought to, the old system can be discontinued, leaving only the new IS up and running. For obvious reasons, this method is often used where the computer IS is critical to the company.

Disadvantages

This changeover method is very expensive. During the parallel changeover, the work of the company's IS is carried out twice. This could cost the company twice as much in staff and resources, because extra staff, computer hardware and software are required to run two systems. Also, staff may well find using two systems confusing and tiring. The timescale to install the new system is necessarily rather long causing a delay in the company realizing its benefits. A Machiavellian view might also be that staff resistance to the new system is given a longer time to demolish the whole thing.

Advantages

The parallel changeover has the built-in safety net of the old system to fall back on should very serious problems arise. Smaller errors can be dealt with without damaging the company because the old system is, for a time, still running the company IS. Perhaps, because it is not so daunting as the direct, big bang method, less time and resources might be needed for its planning, so it may be more economic. It also allows time for the new system to be refined and fine-tuned as users compare the old one with the new.

19.4.3 Phased (staged) changeover

> **Phased installation** Changing from the old information to the new one incrementally, starting with one or a few functional components and then gradually extending the installation to cover the whole new system.
>
> *Hoffer et al., 1999*

This is an incremental or gradual change from the old system to the new one. The system is installed in pieces, stages, or phases.

To do this, the IS must be split up into several self-contained parts or functional components, so when the users are happy with the installation of that part of the IS, another part can be introduced afterwards. The phased method can install each new component either in parallel with the old or directly—in this way it is mid-way between the two previously described changeover methods.

Disadvantages

The old parts of the IS and the new parts need to interact with each other, so bridging programs may need to be produced, costing time and money. Some phased changeovers may fail or be impossible because the old and new systems cannot be linked together in this way. It can become very drawn out, thereby boring and exhausting the staff and developers.

Advantages

The aim of the phased installation is to limit the risk to the company. For the loss or failure of one functional component is not as bad as the failure of the entire system. The benefits of each component may be realized as soon as it is installed. Also, the training for the system can be spread over the changeover period.

> **Single location installation** Trying out a new information system at one site and using the experience to decide if and how the new system should be deployed throughout the organization.
>
> *Hoffer et al., 1999*

19.4.3 Pilot changeover

In this method, the IS is initially installed in one place, such as one branch or one part of the country, and—if it is deemed to be all right—only then installing the system elsewhere. It can also be called the *single location installation*.

The pilot version acts as a trial for *cutting over or rolling out* the IS to the whole company. Naturally, this method is only applicable in rather large companies where the same IS is to be installed in the same way in many parts of the company.

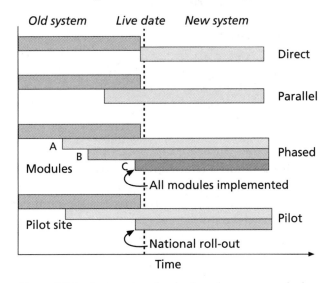

Figure 19.1 A time chart for the four changeover methods.

Disadvantages

The pilot changeover can be a long process, leading to the motivational problems we have outlined above. Also, in the pilot part of the company, extra resources may be required to run the new system and the old one, probably in parallel.

Advantages

The main advantage of this method is that problems with the new system can be discovered in a real-life test, and resolved in the pilot site. Therefore, the potential damage of the failure of the new IS can be limited to only a small part of the company. On the other hand, success at the pilot site can act as very good publicity for the new computer system, boosting confidence in it. This changeover method is probably essential for large, national and multinational organizations.

19.4.4 Overview of the changeover methods

A good way of envisaging the different installation procedures is in the form of a time chart Figure 19.1 shows this.

19.5 Tutorial 19.1

Report design

1. The Video club wants to produce a report showing a detailed monthly sales analysis of its products. It is to contain the following details:

For each video	Category
	Video number
	Video title
	Number of loans of the video during month
	Income generated during month
For each category	Total of loans during month
	Total income generated during month
Grand Totals	Loans during month
	Income during month

Design a report layout to provide the above information.

2. Design a pre-printed membership form for the Video Club. It is to contain:

Name of member
Address
Membership number
Date of joining
Signature
Suitable logo + conditions of membership

19.6 Tutorial 19.2

Screen design

1. Design a character type screen for the Video club, allowing the user to display the current whereabouts of a selected video.
2. Design a character type enquiry screen for the Video club to display all the titles, and their whereabouts for a selected category. The whereabouts are either 'in stock' or the name and address of the user who has it out on loan.
3. The Video club wants a screen program to allow members to book a video or videos by telephone. The program must firstly allow the member's details to be verified – the user may not know their membership number, so a partial access by name is required. They may also not know the video number, only the title. More than one video may also be required.

Design suitable windows type screens for the above program.

20 Documentation

- [ ] The importance of documentation
- [ ] System documentation
- [ ] User documentation
- [ ] Other information
- [ ] Tutorial 20.1

20.1 The importance of documentation

If computer hardware and software is the flesh and blood of the new information system (IS), then the documentation is its soul. IS documentation not only allows developers, managers, and users to see the way the system passes through its various development stages, it also assists them to operate the system effectively. The profession does not provide one single set of documentation, because what is to be included will depend upon the development methodology used and the particular type of system developed. However, at least it can be said that the documentation produced ought to be generated throughout the entire life cycle of the IS.

The documentation of the system is a communication tool between all the interested parties. It provides a historical trace of the whole development of the project as well as instructions about its operation. IS documentation may be split into two parts: *system documentation* and *user documentation*.

Pitfall

It is a big mistake to leave the documentation 'to the end, when we have the time'. The time must be made at the point the documentation ought to be created.

20.2 System documentation

This is the part of the IS documentation particularly directed towards the system's developers. However, users will have been involved intimately with its creation, and may well have seen and understood much of it.

Much of this has been dealt with in many places throughout this book. Every deliverable, from the terms of reference report, through the systems analysis

System documentation
Detailed information about a system's design specifications, its internal workings, and its functionally.

Hoffer et al., 1999

diagrams, to the plans for the system installation, constitute part of the system documentation. They are a portion of the whole system documentation, which is made up of all the documents from all the SDLC phases and stages. You will recall that, although the IS developers have produced most of this, the users and management have had a large part in its generation too.

Amongst others, it should include detailed information about the system's feasibility; the analysis and design documents; program specifications and source code, along with test plans and test reports; the interface designs; and the documents concerning the installation of the system.

20.3 User documentation

This is the documentation primarily for the people who are going to operate the system.

Nowadays, a lot of user documentation is presented online as hypertext documents and help text. However, for many systems, it is still made available on paper, in user references and manuals. Regardless of the way it is presented, good user documentation has certain desirable features:

User documentation
Written or other visual information about an application system, how it works, and how to use it.

Hoffer et al., 1999

- **Structure:** It should be clearly and logically laid out with a contents page, a good index, and section and subsection headings. This follows the structure of any professional report as discussed in Chapter 4. In addition, there might well be chapter overviews and summaries, and help tips.
- **Length:** The documents should be as concise as possible and not too wordy.
- **Style:** Generally, short sentences should be used, with a minimum use of jargon. It has to be written for the ordinary user to understand, and (obviously) with correct spelling and grammar.
- **Appearance:** The document should have a professional look and feel.
- **Layout:** This could include graphics, colour coded sections, illustrations, and icons to help the users understand and remember what they need to do.
- **Audience:** Different parts of the user documentation might well be aimed at different user roles. Not everyone needs to read about everything in the system. These parts need to be carefully directed at the particular users and what they have to do.

20.4 Other information

It may also be useful to include other details in the documentation for the user.

This could be definitions and explanations of **general computer concepts** such as types of software and files, and **IS concepts** like batch and online processing. It might also be useful to write about **organizational features** such as the way the system interfaces with the wider company. Furthermore, you may need to include details of **system management**: the way changes to the system can be requested, as well as the way **system installation** will be achieved.

20.5 Tutorial 20.1

Installation, Documentation & Training

Which installation strategy would you recommend for the Video Shop, giving your reasons?

If you recommend different approaches, explain why.

Compare the four different installation methods, showing their advantages and disadvantages. List these in a table.

Look carefully at the advantages and disadvantages and make a direct comparison when an advantage of one is a disadvantage of the other.

Suppose you were responsible for establishing a training programme for the users of the system in the Video Shop. What user documentation would you produce?

Devise a training plan and discuss how you would implement the training.

Epilogue

We very much hope that this book has been of help to you in your studies. The investigation of information systems (ISs) analysis and design is fundamental to your understanding of computing. Most of you, even the most technologically oriented, will find yourselves working in the field of ISs. For it is ISs that makes most of the money invested in computing courses, students, and lecturers worthwhile.

For those of you who will be, or are already, employed as systems analysts or systems designers, the value of this book will be obvious. However, those of you who are now or are destined to become computer programmers, network or database designers need to understand how your programs, networks and databases will be utilized in the world of ISs.

It should not be forgotten either, that those of you who will not turn out to be computing personnel, will very probably find yourselves being involved in ISs— there are very few professions nowadays (and even fewer in the future) that will not be so engaged. A knowledge of the basics of analysis and design will also be valuable for you.

The book has explored the concepts of systems, systems analysis, and systems analysts. A diagram of the systems development life cycle was followed by requirements analysis. Then we had the very important discussion about writing reports and carrying out presentations. A short discourse on soft systems techniques followed, before the meaty subjects of entity relationship diagrams (ERDs) and data flow diagrams (DFDs). Afterwards, there was a chapter on entity life histories and, briefly, on object oriented techniques. Process descriptions came next and then the rather daunting concepts of logicalization.

This took us into the realms of the design of ISs. The valuable operation of relational data analysis (RDA), also known as normalization, came before the data requirements and the processing requirements of the proposed system. The storage of data in a computer system came afterwards and then a discussion of the interfaces of the system with the user. Then the installation of the new system was included before the book concluded with the topic of the documentation for a computer system.

Finally, we wish you well in your future studies and career. We trust that we have been able to assist you on your path along the highway of information systems analysis and design.

References

Ashworth, C. and Goodland, M. (1990) *SSADM: A Practical Approach*. Maidenhead: McGraw-Hill.

Avison, D. E. and Wood-Harper, A. T. (1990) *Multiview: An Exploration in Information Systems Development*. London: Blackwell.

Bertalanffy, L. von (1968) *General System Theory*. London: Allen Lane.

Bennett, S., McRobb, S., Farmer, R. (1999) *Object-Oriented Systems Analysis and Design using UML*. Maidenhead: McGraw-Hill.

Checkland, P. and Scholes, J. (1990) *Soft Systems Methodology In Action*. Chichester: Wiley.

Checkland, P. (1981) *Systems Thinking Systems Practice*. Chichester: Wiley.

Chester, M. F. (1995) 'Inheritance and classification – a short philosophical investigation', *Third BCS-ISM Conference on Information Systems Methodologies*, pp. 157–1614.

Chester, M. F. (1998) 'The logic of scientific systems development'. *EASE-98, Keele*.

Chester, M. F. (1999) The critical approach to software development, *Information Systems – The Next Generation, 4th UKAIS conference, York*. Maidenhead: McGraw-Hill.

Chester, M. F. (2000) 'CRISP: the campaign for real information systems production'. *ISD2000, Kristiansand, Norway*.

Chester, M. F. and O'Brien, M. (1995) 'Analyst-programmers seen as harmful to software quality'. *Third BCS-ISM Conference on Information Systems Methodologies*, pp. 63–714.

Chester, M. F. and O'Brien, M. (1997) 'Quality added and quality chains'. *International Conference on Software Quality Engineering*, pp. 81–93.

Codd, E. F. (1970) 'A relational model of data for large relational databases'. *Communications of the ACM*, **13**, 77–87.

Connolly, T.M. and Begg, C.E. (1999). Database Approach: a practical approach to design, implementation and management: Addison-Wesley.

Date, C. J. (1975) *An Introduction to Database Systems*, 6th edn. Reading, MA: Addison-Wesley.

De Marco, T. (1979) *Structured Analysis and Systems Specification*. New York: Yourdon Press.

Eva, Malcolm, (1995) *SSADM Version 4: A User's Guide*, 2nd edn. London: McGraw-Hill.

Flynn, D. (1998) *Information System Requirements: Determination & Analysis*. Maidenhead: McGraw-Hill.

Gane, C. P. and Sarson, T. (1979) *Structured Systems Analysis: Tools and Techniques*. Englewood Cliffs, NJ: Prentice-Hall.

Goodland, M. and Slater, C. (1995) *SSADM Version 4: A Practical Approach*. Maidenhead: McGraw-Hill.

Harry, M. J. S. (1997) *Information Systems in Business*, 2nd edn. London: Pitman Publishing.

Hoffer, J. A., George, J. F. and Valacich, J. S. (1999) *Modern Systems Analysis & Design*, 2nd edn, CA: Benjamin/Cummings.

Jackson, M. (1973) *Principles of Program Design*. London: Academic Press.

Jayaratna, N. (1994) *Understanding and Evaluating Methodologies, NIMSAD, a Systematic Framework*. Maidenhead: McGraw-Hill.

Martin, J. (1991) *Rapid Application Development*. New York: Macmillan.

Miles, R. K. (1991) Combining 'soft' and 'hard' systems practice: grafting or embedding? *J. of Appl. Sys. Anal.*, **15**.

Mumford, E. (1983) *Designing Human Systems for New Technology: the ETHICS Method*. Manchester Business School: Manchester.

Patching, D. (1990) *Practical Soft Systems Analysis*. FT Pitman.

Somerville, I. (1989) *Software Engineering*, 3rd edn. Addison Wesley.

Weaver, P. L., Lambrou, N. and Walkley, M. (1998) *Practical SSADM. Version 4+: A Complete Tutorial Guide*. FT Pitman.

Yeates, D. and Cadell, J. ((1996) Project Management for Information Systems. Pitman Publishing.

Yourdon. E. N. (1976) *Techniques of Program Structure and Design*. Englewood Cliffs, NJ: Prentice-Hall.

Index